# The Ecology of a Tallgrass Treasure

## *Audubon's Spring Creek Prairie*

## Paul A. Johnsgard

University of Nebraska–Lincoln

Zea Books, Lincoln, Nebraska: 2018

Copyright © 2018 Paul A. Johnsgard

ISBN 978-1-60962-131-5

doi 10.13014/K25B00NK

Composed in Sitka types.

Zea Books are published by the
University of Nebraska–Lincoln Libraries

Electronic (pdf) edition available online at
http://digitalcommons.unl.edu/zeabook/

Print edition available from
http://www.lulu.com/spotlight/unllib

Spring Creek Prairie Audubon Center
3 miles south of Denton, Nebraska

*Dedicated in memory of my son Jay Erik Johnsgard (1958–2018),*
*whose life was too short by half*

# Abstract

This book describes the major plant and animal components of Spring Creek Prairie Audubon Center, an 850-acre National Audubon Society tallgrass prairie in Lancaster County, southeastern Nebraska. In addition to providing a species list of the area's plants (368 species), there are comprehensive annotated lists of its birds (240), mammals (43), reptiles (23), and amphibians (10). There are also variably complete annotated lists of the area's butterflies (76), sphinx moths (30), silk moths (7), dragonflies (24), damselflies (11), grasshoppers (9), katydids (11), mantids (2), and walkingsticks (2). Brief profiles of life histories and ecologies of 55 animal and 7 plant species are included, as well as information on nearly 100 public-access native grasslands in eastern Nebraska. The text comprises more than 68,000 words, 400 references, and a glossary of 125 biological/scientific terms as well as more than 40 line drawings by the author.

*Correspondence:* pajohnsgard@gmail.com

# Contents

# Figures

# Acknowledgments

I hope this book will provide an insight into the biology and ecology of Spring Creek Prairie and to other relict tallgrass prairies of eastern Nebraska, which represent one of America's most valuable and endangered ecosystems. Little of this information is directly based on my own work, and in particular I have relied heavily on the staff of Spring Creek Prairie Audubon Center. I owe a special debt of gratitude to the late Marian Langan (1962–2017), who was the second director of the center (from 1998 to 2015) and executive director of Audubon Nebraska (from 2011 to 2017) and whose conservation work endures and has inspired us all.

I have also been helped by Kay Kottas, Robert Kaul, and Steve Rolfsmeier with information on the plants of tallgrass prairies, and have benefited greatly from the pioneering work on other Nebraska tallgrass prairies by historic ecologists such as John Weaver, T. L. Steiger, Frederic Clements, and their colleagues. The local Spring Creek Prairie checklists of birds, mammals, herptiles, butterflies, and dragonflies formed the basis for my own checklists, with the help of Kevin Poague and Jason St. Sauver. Don Paseka, Steve Spomer, and Dennis Ferraro provided advice on dragonflies, butterflies, and herptiles, respectively. Two articles I wrote for *Prairie Fire* newspaper (Johnsgard, 2009a, 2012b) are used in my discussion of Spring Creek Prairie in chapters 1 and 2; I appreciate permission from Cris Trautner to reprint them here. Additionally, parts of my discussion on the ecology of Nebraska's tallgrass prairies are largely based on materials that I published in *The Nature of Nebraska: Ecology and Biodiversity*

(Johnsgard, 2001b). The drawings are all my own and in my copyright.

Thanks to the efforts of Dave Sands, Audubon Nebraska, and the National Audubon Society in preserving Spring Creek Prairie, and earlier efforts by University of Nebraska botanists such as A. T. Harrison and Robert Kaul to preserve Nine-Mile Prairie, we can still study the biological complexities and celebrate the aesthetic beauties of native tallgrass prairie in eastern Nebraska. I also thank the Wachiska Chapter of the National Audubon Society, which is based in Lincoln and serves southeastern Nebraska, for the members' efforts in preserving relict prairies.

The long-term work by Ernie Rousek and Tim Knott has been invaluable in locating and helping to obtain conservation easements on more than 30 surviving tallgrass prairie remnants in southeastern Nebraska. Tim also provided me some information on other relict tallgrass prairie tracts in eastern Nebraska, and Ernie added some historic information on Nine-Mile Prairie. Their and others' work in searching out prairie remnants, facilitating the development of conservation easements, and raising money to purchase and maintain a diversity of tallgrass prairies in southeastern Nebraska has been critical to prairie conservation and education. I also appreciate the use of their descriptions of the Wachiska prairies and directions for reaching them, which I have modified to varying degrees.

Linda Brown accompanied me on many trips to Spring Creek Prairie and other regional prairies, and provided compass corrections when my writing was prone to diverge from the primary subject. Rachel Simpson kindly looked at my lists of plants and corrected spelling errors as well as other writing weaknesses. Several other people have also offered suggestions for or corrections to the text, or helped in other ways, including Scott Johnsgard, Josef Kren, and John Carlini. Finally, I have a boundless debt of gratitude and extend my continuing personal thanks to the University of Nebraska's indefatigable DigitalCommons coordinator, Paul Royster, and to his priceless editorial assistant, Linnea Fredrickson, who have cheerfully tolerated my writing weaknesses for more than a decade. To paraphrase Reinhold Niebuhr: Nothing we do, however well intentioned, can be accomplished alone; therefore, we are saved by great editors.

Paul A. Johnsgard
Lincoln, Nebraska

# ☘ Introduction to Spring Creek Prairie

When I was very young, I used to walk along railroad track rights-of-way near my home in the Red River Valley of eastern North Dakota. I didn't know that the "turkey-foot grass" that grew higher than my head was something special or that under its more formal name of big bluestem it is a charter member of the tallgrass prairie that once covered much of eastern North Dakota. Somewhat later my mother began to teach me some of the native prairie flowers that grew in low meadows near their once-homesteaded farm near the Sheyenne River. Today this area has been preserved as part of the Sheyenne National Grassland, the largest federally owned area of tallgrass prairie in America. I learned there to identify such beautiful plants as tall blazing star and Canada goldenrod, and acquired at least a nodding acquaintance with milkweeds, sunflowers, and some of the other more common and colorful wildflowers. At least as importantly, I learned to associate such glorious birds as marbled godwits and bobolinks with patches of native prairie, which even then were mostly confined to hilly, rocky, or sandy sites at the edges of what was once glacial Lake Agassiz, and now if the heart of the Red River Valley. I much later learned that such relatively rare prairie plants and animals are "indicator species," and that if one wishes to find them (and protect them), it is necessary to protect the entire prairie community.

When I arrived in Lincoln in 1961, there were still dozens of relict prairies near town, where I could go to see the prairie birds and plants of my childhood. But, as the years passed, these prairies disappeared one by one to agriculture or suburban developments. A few scattered memories were often all that remained, in ditches and at the edges of fields where the deep roots of perennial grasses like big bluestems allowed them

*Fig. 1. Lark sparrow on a red cedar branch*

to continue for a time their losing battle against plows and herbicides. Even the fairly new house we bought in 1963 at the then-edge of Lincoln still had a few shoots of big bluestem that fought valiantly for a few years against the socially acceptable Kentucky bluegrass, a semi-weedy European water-hungry and fertilizer-dependent import. After being warned by city authorities about tolerating such "weeds" as bluestem in my yard, I, too, accepted defeat and let natural selection take over.

One of the few remaining public-access prairies near Lincoln persisting into the 1990s was Nine-Mile Prairie, a once privately owned pasture on hilly glacial moraine of nearly 900 acres, which had been studied intensively during the late 1920s by T. L. Steiger, who reported 345 plant species there (Steiger, 1930). During World War II, part of the prairie was taken over by the military for use as an ammunition storage site, and its size gradually diminished to its present 230 acres. The prairie was acquired by the University of Nebraska Foundation in 1983 and is now protected and managed for both research and as a historic prairie. A survey produced a total of 298 vascular plant species (Kaul and Rolfsmeier, 1987). It is freely available to the public for nonconsumptive purposes, such as nature study, birding, and hiking.

By a stroke of good fortune, and some ambitious money-raising on the part of the state and local chapters of the National Audubon Society, Spring Creek Prairie was acquired in 1998. The prairie is three miles south of Denton, or 18 miles south of Lincoln. This prairie is located on unplowed glacial moraine, with a vascular plant inventory of about 350 species, based on the studies of Kay Kottas (2000, 2001). About 80 percent of its original 610 acres is covered with tallgrass prairie; more recent land acquisitions have brought the total area to 850 acres as of 2018. The remaining 20 percent of the original acreage is a combination of wetlands (63 plant species) and woodland dominated by bur oak, green ash, and several elms (62 plant species), while the recent acquisitions are under restoration management toward prairie.

Among the grasses, the five species Kottas found to have the highest frequencies of occurrence at Spring Creek were, in descending number, big bluestem, little bluestem, side-oats grama, Indiangrass, and tall dropseed, all warm-season grasses. At Nine-Mile Prairie the corresponding sequence was big bluestem, Indiangrass, little bluestem, tall dropseed, and side-oats grama. At Spring Creek Prairie the three most abundant cool-season grasses were Kentucky bluegrass, smooth brome, and Scribner's panicum, the first two of which are introduced and invariably invasive species. The incidence of introduced species was found by Kottas (2000, 2001) to comprise 22 percent of the total vascular plant flora at Spring Creek, and 15 percent at Nine-Mile Prairie. The flora of Spring Creek has been further inventoried since 2000, and as of 2018 included more than 370 vascular plant species. The checklist in chapter 21 of 368 species of eastern Nebraska tallgrass prairie plants is based largely on the Spring Creek Prairie list. Of these, about 42 percent are summer flowering, 25 percent are fall flowering, and 9 percent are spring flowering. An additional 10 percent bloom from summer to fall, 9 percent from spring to summer, and 8 percent from spring to fall. Fifty-four percent of the total taxa are perennials, 43 percent are annuals, and 3 percent are biennials.

Excluding trees, woods-adapted forbs, and wetland plants, there are about 225 upland prairie species at Spring Creek. Although the vast majority of the vegetational coverage consists of perennial grasses, grass species make up only about 20 percent of the taxonomic diversity. Forbs (herbaceous plants other than grasses and sedges) compose about 70 percent. Additional species, and shrubs plus a few woody vines make up the remaining 8 percent. So, it is the forbs that give the greatest structural complexity to tallgrass prairie, and these include a large number of plants in the sunflower (aster) family, fewer in the legume and grass families. Some families are represented by very few species. For example, only one orchid species (nodding ladies' tresses) has been found at Spring Creek, and four at Nine-Mile Prairie. Of the 20 most abundant forbs (broad-leaved herbs) present in Spring Creek and Nine-Mile Prairies, 13 were found by Kottas to be present at both prairies. Several of these are introduced and invasive species.

Although the grasses are all wind pollinated, many of the forbs are pollinated by insects, and it is this latter adaptation that has produced the displays of multicolored and scented flowers in spring and summer that at times turns the tallgrass prairie into a natural garden. At Spring Creek Prairie the purples of vetches, wood sorrels, and violets set the stage for June's flowering of purple coneflowers and blue ruellias. The black-eyed susans and sunflowers in July are golden summer highlights, and are followed in August and September by a succession of other yellow composite flowers scattered among the then-towering grasses. Several blue to purple asters such as New England and azure asters vie with downy gentians to be the final colorful fall hosts to honeybees and bumblebees until late October.

The animal list for Spring Creek Prairie currently (2018) includes at least 235 bird species plus at least 30 mammals, 54 butterflies, and countless other invertebrates. Spring is my favorite time to visit Spring Creek, when the migratory birds are returning and the first spring flowers such as violets rush into bloom

to complete their flowering before being shaded out by the taller grasses and flowering forbs that attract insects. But each season has its attractions. The tallest prairie grasses are nearly all "warm-season" species, which wait for the oppressive heat of midsummer to put on their most rapid growth. By September, the Indiangrass and big bluestem may easily reach eight feet in a wet year, and to lie down in a stand of these grasses and look to the sky above is to know how an ant's view of its gigantic world might appear.

Spring Creek Prairie Audubon Center is now the jewel in Nebraska's prairie crown, with a new interpretive center and a staff including several trained biologists. Wachiska Audubon Society, the local Wachiska Audubon chapter, donated an additional 14-acre adjacent woodland. Later an adjoining 168-acre parcel of prairie was purchased, made possible with a major gift from the Charles R. McConnell family. Another addition has brought the property's total acreage to 850 acres. The latter two acquisitions are undergoing prairie renovation by controlled burning, mowing, and other techniques, which allows visitors to see how overgrazed prairies can be restored to near-pristine condition.

The tallest hills of Spring Creek Prairie are among the highest points in Lancaster County, affording a spectacular unobstructed view in all directions. Sitting on a hilltop, one can close one's eyes and listen to the sounds of near solitude, sometimes marked only by the songs of a distant meadowlark, the scream of a soaring red-tailed hawk, or in spring the soft kettledrum sounds of courting greater prairie-chickens. A few years ago, the autumnal equinox happened to fall almost exactly on the night of the full moon, so I decided to watch the simultaneous sunset and moonrise from the top of one of these tall hills. I sat on a large quartzite boulder that protruded a few feet above the ground, a souvenir from the melting glacier that had shaped these hills. It was like watching one beautiful curtain fall in the west as another equally stunning curtain rose in the east. As dusk slowly transformed into darkness, some of the last fireflies of the year began their nightly performance, as if their light might make the autumnal warmth last just a little longer. By November, the prairie has quieted down, with the starches, sugars, and other carbohydrates that were manufactured by perennial plants during summer safely stored in root systems many feet below ground, well out of reach of grazing animals. Left behind are the rusty brown and golden skeletons of leaves and stems that make for spectacular fall panoramas, especially when contrasted with the red leaves of sumacs and the cerulean hues of cloudless fall skies.

Winter is a time for hardy souls to walk the prairie trails in search of snowy tracks marking the passage of coyotes, rabbits, deer, raccoons, mice, and other mammals whose presence would otherwise likely remain undetected. Much of the activity of small rodents occurs under the snow; its insulating quality allows the temperature at ground level to remain only a few degrees below freezing, even if the air temperature above the snow should approach zero. Foxes, coyotes, and some owls can hear the sounds made by unseen mice and voles as they scurry about below, and the predators will suddenly pounce on them from above. By December, the long blue shadows of grass cast on the snow by the pallid winter sun provide only cold comfort to the visitor but do offer the promise of a sun that by January will be rising sooner, slowly increasing in strength, and providing both life-giving light and heat to the waiting plants and animals.

*Native prairie is old. It has existed since the stars sang together, yet stays as young as April. Each year its grasses and forbs renew themselves, flowering and fruiting and retreating for winter into the deep banks of fertility they have fostered, always returning in the spring.*

John Madson, *Tallgrass Prairie*

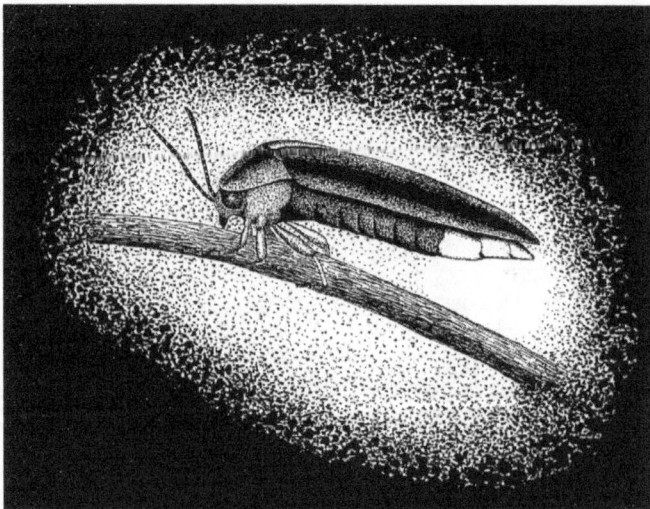

*Fig. 2. Firefly*

# 2 The Educational Values of Spring Creek Prairie

Spring Creek Prairie shimmers like a newly woven copper-colored tapestry in the brilliant sunlight of late October. Its 850 acres of glacially sculpted land in southeastern Nebraska are located less than 20 miles southwest of Lincoln, its high hills representing the western limits of the only great glacier's edge that reached this far south. Spring Creek's hilly ground is intermixed with rich soil materials thus carried in from the north and blown in from the west, but its undulating surface and rock-strewn substrate have protected it from the plowing and cropping that were the fate of nearly all of eastern Nebraska's most fertile lands.

Many scattered and rounded quartzite boulders, transported by ice from as far north as South Dakota and Minnesota, are visible near the tops of some of the higher hills, where they were de-posited during the melting of the Kansan glaciation. Nearly a million years of subsequent erosion have gradually exposed some of them. Yet even the largest of these boulders become seasonally hidden by the tall perennial grasses after they have attained maturity in early fall. The tallest of the grasses, big bluestem and Indiangrass, grow as high as eight to ten feet after generous summer rains. These towering grasses effectively hide most of the other nearly 370 species of plants that have been identified in this relict prairie preserve, one of the largest remaining stands of tallgrass prairies in Nebraska.

Because of its diverse plant and animal life, this prairie has been a magnet for naturalists ever since the National Audubon Society acquired it in 1998. Its educational value was greatly enhanced with the completion of a beautiful inter-

*Fig. 3. Red-winged blackbird, male*

pretive center in 2006. With its hay-bale construction and low-slung architecture, the center fits appropriately and inconspicuously into its prairie surroundings.

Over the past decade Spring Creek Prairie Audubon Center and Lincoln's Pioneers Park Nature Center have expanded their educational programs to encompass all of the children in Lincoln's fourth-grade public schools during the fall season. During spring, fourth-graders from other towns within a roughly 50-mile radius are also brought in to spend most of a day on the prairie. As a result, over the years several thousand students have been able to see the prairie firsthand, and under the guidance of preserve staff and volunteers learn about both its natural history and human history.

There are also other seasonal educational opportunities for both children and adults at Spring Creek, such as participating in periodic nature and birding walks, or attending early fall festivals such as Twilight on the Tallgrass and Harvest of Traditions. During a recent prairie festival, visitors could walk the Prairie Appreciation Trail and stop at sites to catch and identify live grassland insects, capture and examine aquatic invertebrates from one of the wetlands, draw or color prairie plants and birds, or learn about some of Nebraska's endangered species.

A few years ago, I accompanied nearly 150 students from Lincoln's Roper Elementary School as they ventured out on the prairie in late October, most of them probably having never before set foot on the grasslands that were so familiar and vital to earlier generations of Nebraskans. Divided into small groups of about a dozen students for each of their teachers and a Spring Creek mentor, the kids were soon literally immersed in the tall grasses, some of which were twice as tall as they were.

At an early stop the children learned of the complex grassland composition of tallgrass prairie and its hundreds of kinds of grasses and other plants, many of them wildflowers still in bloom. There were also populations of bees and other insects, trading pollination for bounty in the form of nectar, pollen, or leaves.

The students also learned of the poisonous juices of the common milkweed. These chemicals protect it from being eaten by nearly all insects except the larvae of monarch butterflies and milkweed beetles, both of which have developed chemical methods for neutralizing the lethal effects. The children also enjoyed learning about the dispersal abilities of milkweeds by releasing their parachute-like seeds into the air and watching them drift away in the breeze.

Another early stop was at the bottom of a long, gently sloping hill, marked by depressions that are the remnants of pioneers' wagon ruts formed during the mid-1800s. At that time wagon trains from Nebraska City cut through the then-virgin grasslands and passed directly through what is now Spring Creek Prairie. Many were carrying people and supplies to Fort Kearny—about 100 miles west along the Platte River, an important stopover on the historic Platte River Road—or California or Oregon, more than 1,200 miles and many unknown mountain ranges westward. These travelers were the harbingers of a flood of at least 300,000 people whose dreams would be jam-packed into wagonloads of wooden boxes and trunks but many of whom would soon encounter a nightmare of accidents, illnesses like cholera, and other unforeseen dangers.

From the bottom of the imposing hill the kids worked their way slowly upward, sometimes stopping to look at grasshoppers or other insects, or to gaze at the seemingly limitless expanse of prairie all around them. About half-way up the hill they assembled on a grassy slope, where they were asked to sit, relax, and close their eyes. They then were told to imagine that the time was the 1860s, and that each of them was a badger, living in a burrow, and surviving amidst the quiet of the prairie landscape. They were next asked to imagine hearing the sounds of a wagon train approaching for the very first time, and wonder how from that time onward their lives would be forever changed, as would those of all other prairie inhabitants and, eventually, the prairies themselves.

Later, small groups of the children used nets to capture insects and, with the help of mentors, learn the basics of identifying them. Others used plastic hoops to mark a small circular patch of prairie and then closely examined the variety of plants that were within it. Still others used blank postcards to draw on, and to color them, using natural pigments squeezed from the berries of prairie plants.

After two hours on the prairie, the kids returned to the education building for picnic lunches. They then returned to the prairie for a brief afternoon session before taking a school bus back to Lincoln, tired but filled with memories likely to endure for a lifetime.

It is impossible to know if these unique experiences permanently alter the children's perceptions of Nebraska's prairie heritage, but letters received later by Spring Creek Prairie staff provide some evidence. One of them reads: "Thank you for teaching me about the prairie. I learned about insects, different kinds of plants, and grasses. I liked drawing pictures with berries, flowers, and leaves. I also enjoyed catching insects with a butterfly net and a jar. Now I want to go there again with my family, to see many insects and animals! Your friend, Hee Jo."

I hope that not only Hee Jo but also many of the other children will indeed return to the prairie again and again. It will help them become emotionally attached to the natural world and to recognize the value of preserving habitats such as Spring Creek Prairie. The prairie is an ever-changing and ever-fascinating tapestry of life, teaching important ecological lessons not so easily found in books or so willingly assimilated. It is one of Nebraska's great natural treasures that, like the Missouri and Platte Rivers, Chimney Rock and Scotts Bluff, provide visual reminders of who we are as a state and nation, and quite literally where we came from. It should also remind us of the importance of preserving and protecting all these rare icons of historic America, which, like our freedoms from governmental tyranny, are now as much uncertain as were the fates of the brave pioneers crossing the prairies some 160 years ago.

*Even small tracts of prairie, although surrounded by pastures with their cultivated fields with their accompanying annual and long lived weeds, remain intact. Such stability denotes a high degree of equilibrium between native vegetation and its environment under the control of a grassland climate.*

J. E. Weaver, *Native Vegetation of Nebraska*

# 3 The Ecology and Behavior of Tallgrass Prairie Birds

*And so I turn my eyes toward the east and west each spring, making certain that I witness at least one sunrise and sunset in the company of grassland birds.*

Paul Johnsgard, *Prairie Birds: Fragile Splendor in the Great Plains*

There are 448 land breeding bird species in North America, 192 of which were classified by Partners in Flight (Rich et al., 2004) as being Species of Continental Importance from a conservation perspective. Of this group, nearly 40 percent breed within the Prairie Avifaunal Biome, as defined by these authors. The Prairie Avifaunal Biome exists in a region that encompasses the Great Plains and the interior lowlands of the upper Mississippi River Basin. This area includes all of the grasslands east of the Rocky Mountains, from the western shortgrass steppes through the central mixed-grass prairies to the eastern tallgrass prairies. By the late 1990s all but one of the 40 Prairie Avifaunal Biome's bird species had population trends that were declining or were of uncertain trend status.

Of all the 320 species that were judged to be breeding within the Great Plains states (excluding Texas) as of the late 1970s (Johnsgard, 1979), I later defined 32 as "grassland endemics, whose breeding ecologies are clearly associated with grasslands, and whose breeding ranges show strong affinities with the Great Plains" (Johnsgard, 2001a). Although some of the Great Plains' breeding birds are permanent resident species, many more migrate to winter in the southern United States and south into Central and South America. The arid highlands of northwestern Mexico were the ancestral home of much of the Great Plains flora, which slowly expanded northward with the rise of the Rocky Mountains in early Cenozoic times, or less than 50 million years ago. The resultant increasing "rain-shadow" aridity on the eastward (lee) side of the mountains that was generated by this mountain-building trend probably allowed drought-resistant plants, such as rapidly growing annual and perennial grasses, to extend their range northward. Ground-foraging and seed-eating passerine birds, such as sparrows and finches, as well as larger seedeaters such as quails and doves, likewise expanded northward into the developing Great Plains (Johnsgard, 2001a).

The long-term result of these range and climatic changes was the evolution of a guild of birds whose ecological niches became increasingly adapted to seed-eating. Their seasonal breeding cycles became tied to the region's relatively short summer growing seasons, when grasses and forbs are growing most rapidly and when plant-eating insects are most abundant and provide food for young birds. The birds' legs and feet became increasingly adapted to walking, running, and ground scratching, and their beaks modified for seed crushing, together with other behavioral, morphological, and anatomical adaptations (Fig. 4).

The low vegetation of a grassland environment also made a brown, camouflaged plumage a highly adaptive trait. For prairie birds, conspicuous colorful or contrasting white-patterned markings became restricted to body locations

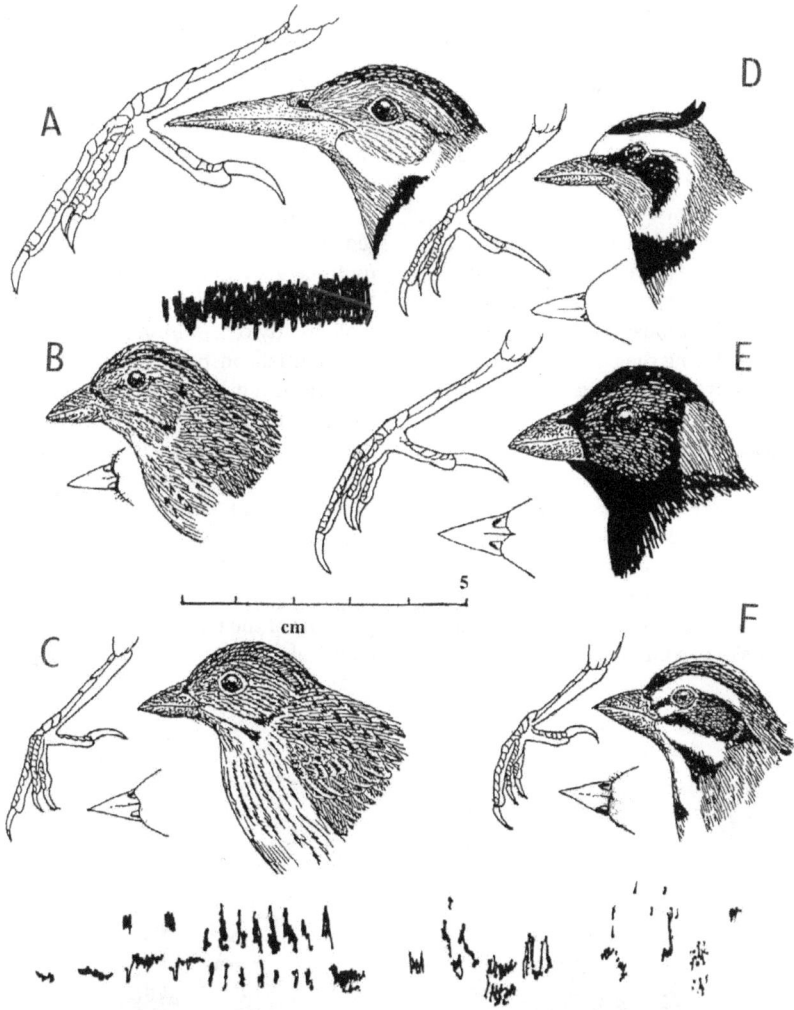

*Fig. 4. Heads and feet of (A) eastern meadowlark, (B) grasshopper sparrow, (C) vesper sparrow, (D) horned lark, (E) bobolink, and (F) lark sparrow. Song sonograms are shown for B, C, and F.*

where they could usually be kept concealed, such as on the outer tail feathers, the upper wing-coverts, or the under tail-coverts. However, they can suddenly become visible in flight, by wing-raising and wing-flapping, and by tail-spreading or tail-cocking, or by expanding colorful areas of their skin that are normally hidden, the way prairie-chickens do. Males of many prairie-adapted birds often have complex songs, which are used to define and defend territories, and to attract same-species females. These "advertising" songs are often uttered, and the bird's importance proclaimed, from shrubs, fence posts, telephone poles, or

other elevated sites within a male's territory. More rarely, the songs are deliv-
ered from the sky during song-flights,
when conspicuous wing and tail patterns
can also be effectively exhibited. Meadowlarks, horned larks, bobolinks, long-
spurs, upland sandpipers, and long-billed
curlews are notable among the grassland birds for their sometimes spectacular song-flights. Even such ground-loving
species as greater prairie-chickens and
sharp-tailed grouse perform brief aerial
displays during their otherwise ground-
based courtship activities.

Of the 32 species defined by Johnsgard (2001b) as grassland endemics,
at least 26 are regular breeders in Nebraska, and 10 of these probably breed
at Spring Creek Prairie. Among the most
common of these are two of the most
typical North American grassland birds,
the eastern and western meadowlarks.
Nebraska is somewhat unusual in that

over some parts of the state both eastern and western meadowlarks can be
seen and heard at the same location.
Where they commonly occur together,
as in the eastern fourth of the state, the
eastern meadowlark is likely to be found
in the lower, moister sites, and the western on uplands and drier habitats. Often both can be heard singing almost simultaneously, and it is the difference in
the advertising songs of the males that
make ready recognition and avoidance
of potential conspecific mates possible.
The western has a more extended, complex, and melodious series of many short
notes, uttered too rapidly to count. The
eastern has a more trombone-like series
of fewer, sliding-scale notes (Fig. 5). Intermediate songs are sometimes present
and may leave the observer in doubt as
to the singer's identity, as the plumage
patterns of the meadowlarks are nearly
identical.

Fig. 5. Sketches and song sonograms of male eastern meadowlark (left)
and western meadowlark (right)

Individual male western meadow-larks sing a variety of unique song types, usually ranging from about 3 to 12. Some of these song types may be shared with other males in the local population, but no two males exhibit exactly the same repertoire. A male may repeat one of his song types several times but will switch to a different type upon hearing a rival, perhaps to reduce the likelihood of this other male becoming less responsive to a particular song type. Song-switching may also be important both in territorial defense and in achieving mate attraction. Males having the largest song repertoires also tend to be among the first to obtain mates, and have greater reproductive success than do less versatile males (Lanyon, 1994), suggesting that song is one of the most effective ways to attract a female, as humans also much more recently discovered (Darwin, 1871).

Two other species with song-flights and extended songs are the bobolink and the horned lark. The bobolink (Fig. 6) is a species with a high degree of sexual dimorphism. Their song-flights rarely rise more than about ten feet above ground but are especially conspicuous because of the male's eye-catching combination of black, white, and gold. By comparison, the monogamous horned lark (Fig. 7) sings while flying much higher in the sky, and it performs a much longer song-flight. In the horned lark the two sexes have identical plumage, further setting it apart from the bobolink. Other monogamous grassland birds, such as the grasshopper sparrow (Fig. 4) and Henslow's sparrow (Fig. 6), also have rather short songs that are not highly complex acoustically and are relatively low in volume. In contrast, male dickcissels (Fig. 6) are persistent and loud singers, and might be considered semipolygynous. Especially vigorous dickcissel males are sometime able to acquire a second mate; these later second matings are more likely to escape brood parasitism by female cowbirds, which usually stop laying by mid-July.

A higher development of a polygynous mating system is practiced by the bobolink, in which a single male might have as many as four mates. The male plumage of the bobolink is totally different from, and much more conspicuous than, that of the sparrows and horned lark. Such interspecies differences in sexual dimorphism can be attributed to the effects of sexual selection in nonmonogamous birds. In these birds, a single male might be able to fertilize several females during a single breeding season, especially if his appearance and sexual-attraction behaviors allow him to out-compete other males and represent a reliable indication to females of his better genes and higher vigor. These polygynous males contribute a greater genetic input to the overall population than do less genetically fit males and favor the evolution of increased sexual dimorphism in both their plumage and behavior (Darwin, 1871). A promiscuous mating system is even more effective in promoting sexual selection, such as in the lek-breeding system of greater prairie-chickens and other grassland-breeding grouse. In such systems a single dominant male might be able to fertilize 80 to 90 percent of the females within a local population (Johnsgard, 2001b).

As noted earlier, national populations of the grasslands birds of North America have suffered greatly from the conversion of native prairies into agricultural fields. This trend has been especially evident in the agriculturally productive Great Plains where, among the 32 species identified in 2001 as grassland endemics, only one (the ferruginous hawk) exhibited a highly significant long-term increasing population trend during the 1966–1996 Breeding Bird Surveys. In contrast, six of these species exhibited highly significant population declines

*Fig. 6. Sketches and song sonograms of male bobolink (above),*
*dickcissel (lower left), and Henslow's sparrow (lower right)*

over that entire period; four others had statistically less significant population declines. One species (the grasshopper sparrow) had a highly significant decline for only the more recent half of that survey period (Johnsgard, 2001b). Mineau and Whiteside (2013) concluded that acute toxicity from pesticides, rather than agricultural intensification, has been the primary cause of long-term population declines in North American grassland birds. They judged that

*Fig. 7. Sketches and song sonogram of male horned lark*

the largest number of declining species has been in Minnesota, with 12 grassland species affected, followed by Wisconsin, with 11 species, and a group of 5 species, including in Nebraska, where 11 species have been declining.

In a review of Breeding Bird Survey data for the period 1966 to 2011 (Sauer et al., 2013), the 41 total species grassland birds they analyzed exhibited consistent declines throughout the interval, and 24 obligate grassland species collectively declined 37.8 percent. Bringing the trend data more up to date among the 10 grassland endemics breeding at Spring Creek Prairie, only the horned lark and greater prairie-chicken exhibited possible small long-term population increases during the period 1966–2015 (Table 1). All the other 8 species underwent appar-ent downward population trends, and their estimates were classified as reaching levels of "moderate" statistical precision. Relative breeding abundance estimates among 11 grassland species in shortgrass, mixed-grass, and tallgrass Breeding Bird Surveys between 1966 and the mid-1990s are also shown in Table 1.

**Table 1.** Occurrence Frequencies of 11 Prairie Birds in Three Grassland Habitats, Based on Breeding Bird Censuses, 1972–1996[a]

|  | Tallgrass 52 routes | Mixed-grass 120 routes | Shortgrass 9 routes |
|---|---|---|---|
| Dickcissel (−0.36%)[d] | 83% | 1% | — |
| Eastern meadowlark (−3.28%)[d] | 50% | 8% | — |
| Grasshopper sparrow (−2.52%)[d] | 61% | 58% | 22% |
| Western meadowlark (−1.29%)[d] | 36% | 93% | 33% |
| Bobolink (−2.06%)[d] | 21% | 55% | — |
| Upland sandpiper (−2.46%)[d] | 21% | 46% | — |
| Henslow's sparrow (−1.53%)[b] | 8% | — | — |
| Greater prairie–chicken (+2.8%)[c] | 6% | — | — |
| Vesper sparrow (−0.85%)[d] | 4% | 3% | — |
| Horned lark (+0.40%)[d] | 2% | 7% | 78% |
| Lark sparrow (−0.78%)[d] | 2% | 1% | — |

a. After data of Johnsgard (2001a). Figures in parentheses indicate estimated long-term annual population trends based on Breeding Bird Survey routes for the period 1966–2015.

b. Estimate based on less than 5 long-term routes, with very low abundance or imprecise results.

c. Estimate based on less than 14 long-term routes, with low abundance or imprecise results.

d. Estimate based on at least 14 long-term routes, with moderate precision and route abundance.

# ④ The Ominous Future of the Greater Prairie-chicken

The greater prairie-chicken has an English vernacular name that sadly understates both its beauty and aesthetic value. Granted this name makes clear that the bird's presence provides a reliable indication of native prairies, and it is somewhat "greater" in size than the lesser prairie-chicken, which was eliminated from western Nebraska more than 70 years ago. But the prairie-chicken is no more a chicken than a turkey is from Turkey. Perhaps the prairie-chicken should have been called something like "soul-of-the prairie" or "spirit-of-the-grasslands" to force those who want to kill it to think twice about their motives. Anyone who has spent a spring sunrise with prairie-chickens will know exactly what is meant by these semantic intima-

tions of the holy; there is a sense of the sublime when one is in the presence of displaying prairie-chickens. They are acting out the identical courtship routines they inherited from distant ancestors, on grassland sites made sacred through their annual use by uncountable generations past. Additionally they are determining, by both battle and bluff, which individual males will transmit their genes to the next generation by being differentially able to attract the majority of the females that visit the lek and are ready to lay their eggs.

Darwin's concept of survival and reproduction of the fittest is played out on a daily basis on grassy hilltops (leks) every spring. Being able to witness these performances is an auspicious act in

*Fig. 8. Greater prairie-chicken males fighting*

the original Latin sense; the actions of the birds provide a reliable augury relative to the future fortunes of the species. Prairie-chicken mating displays represent a complex mixture of tail-raising, wing-drooping, foot-stamping, and low-pitched calling (booming) that is performed with the sides of the male's neck inflated to the size of small oranges. There are also male-to-male threats and violent fights (Fig. 8), as well as subservient displays performed only to females, as invitations to copulation. In any assemblage of courting males, a single male will, by threats and attacks, be able to dominate all the others. Such "master cocks" thus gain the right of mating with females without challenges by the other subordinate males, and eventually will inseminate nearly all the females in a local population.

We have far too few sacred natural sites left in eastern Nebraska; most of the Pawnee, Omaha, and Otoe holy sites have since been cleared and "developed," or their exact locations have been long forgotten. But we must not forget the locations of prairie-chicken leks; they whisper to us of secret places where grama grasses and bluestems grow thick on the ground, and where flint arrowheads are likely to lie buried beneath the thatch and loess. They tell us of meadowlark and dickcissel song perches and traditional coyote hunting grounds. They are as much a connection to our past as are the ruts left in the Nebraska soil by immigrant wagons or the preserved raiments of Native American Plains cultures carefully stored in museums.

However, eroding pioneer wagon trails and fading garments are essentially static and retrospective icons; prairie-chickens are the vital essence of life itself, clinging to their brief moments in the sun with all the energy they can muster. They risk attack by both early-rising hawks and late-flying owls simply to have a chance to reproduce before predators, the plow, or a hunter's gun all too quickly eliminates them. Their feathers, which are sometimes strewn over the ground when a predator has been successful, are the camouflage colors of dead grass, and their soft hypnotic voices are both exciting and yet soothing, like the mantras emanating from a Hindu temple. The birds compose a New World symphony all by themselves, a harmony of sound, color, and movement.

It is easy enough to save these wonderful sights and sounds for following generations. We need only to recognize that both prairies and prairie-chickens must be preserved if for no other reason then to help us understand what Nebraskans such as Willa Cather meant when she wrote lovingly of our "shaggy grass land," or what John Weaver meant when he said that "civilized man is destroying a masterpiece of nature without recording for posterity that which he has destroyed." We may well sometimes destroy the things we love out of ignorance; we should never do it purposefully.

To provide some sense of what is at stake, the eastern race of the greater prairie-chicken (the Pilgrims' famous "heath hen") is now extinct in all seven Atlantic coast states where it once occurred. The last individual heath hens probably died during the year of my birth (1931). The one-time coastal Texas and Louisiana race (Attwater's) was on the verge of extinction by 2018. The lesser prairie-chicken of the arid southern shortgrass plains is probably also doomed because it was reduced by 2018 to less than 35,000 birds, ironically having been removed in 2011 from the nationally threatened list by Texas oil interests. Its population is now considered to be in danger of a "death spiral." The coastal Attwater's race of the greater prairie-chicken was extirpated from Louisiana by 1919. In 1937 hunting in Texas was terminated, and it was classified as endangered in 1957 and critically endangered by 2018. It now is essentially limited to a single national wildlife refuge (Attwater Prairie Chicken National Wildlife Refuge) where, in an emergency life-support situation, only captive breeding and releases of hand-reared birds can maintain its numbers.

By 2018 the greater prairie-chicken's interior race of the Great Plains had been extirpated from 11 states and Canadian provinces, and was marginally surviving in 8 others. Nebraska is now one of only five states where autumn hunting of prairie-chickens was still permitted as of 2015, and in two of these (Minnesota and Colorado) hunters were allowed only two birds per season (Johnsgard, 2016). Three states (Nebraska, Kansas, and South Dakota) still allowed regular, relatively uncontrolled prairie-chicken hunting in 2018, although in none of these states are agency biologists willing to even hazard a guess as to how many birds still exist in their states. The annual hunter kill in Kansas dropped about 90 percent from a peak of 109,000 in the early 1980s to about 12,000 in 1999, paralleling a comparable 70 to 90 percent decline in Missouri's prairie-chicken population over the same time span, and its virtual extirpation by 2018.

Although as recently as 1980 there may have been as many as a million greater prairie-chickens present in North America, by the late 1990s no more than 200,000 to 300,000 were believed to be in existence (Johnsgard, 2001b). Breeding Bird Surveys from 1966 to 2015 indicate that this species' long-term population underwent an average annual decline of 2.8 percent during that period. Among all six states providing survey data, only Nebraska, South Dakota, and Kansas reported apparent increases that were considered to be moderately reliable statistically.

Rather than permitting the killing of prairie-chickens, state game agencies across the Midwest could better spend their energy in preserving every last shred of prairie they can locate, thereby conserving not only prairie-chickens but also more than 200 species of native prairie plants, some 30-odd grassland-adapted birds that, like the prairie-chicken, are nearly all declining nationally, and countless other living things. They would also thus be preserving special places of spiritual renewal for human visitors who prefer seeing, enjoying, and remembering the natural world, rather than providing hunters with opportunities for collecting a few soon-forgotten feathered trophies.

By comparison, a spring sunrise spent in the company of prairie-chickens can be as meaningful as witnessing a miraculous rebirth, for that is what is actually occurring. The odor of freshly greening grass is infinitely more memorable than the stringent smell of burned gunpowder, and the harmonic cooing notes of a dozen male grouse calling simultaneously on a prairie hilltop lek in the half-light of dawn is as compellingly beautiful as a string ensemble playing a late Beethoven quartet. For those who know the current perilous status of prairie-chickens, the soft sounds might also bring to mind the ineffable sadness of the ending of Tchaikovsky's heart-rending Pathétique Symphony, with its intimations of despair and forebodings of death. Perhaps it is not too late to alter this ending, or at least to add a final triumphant coda. We need only to save the prairies to accomplish this small miracle ourselves.

*The face of the earth is a graveyard, and so it has always been.*

Paul Sears, *Deserts on the March*

# 5 The Birds of Spring Creek Prairie

This list of about 240 bird species from Spring Creek Prairie is based on the observations of Audubon staff and others up to May 2017 plus eBird data from 1900 through April 2018. Species shown in **bold** are presumptive or known breeders at Spring Creek Prairie, based on data from there and other local tallgrass sites. Species classified by Johnsgard (2001b) as grassland endemics are shown in *italics*. Indicated habitat preferences are only general and refer to the seasons when the species occurs locally. Months of reported presence are shown for seasonal residents and migrants but not for "vagrants" (those species that have been reported only once). The taxonomic sequence and nomenclature follow the most recent (2018) American Ornithological Society Checklist of North and Middle American Birds, but aquatic and shoreline species are sublisted together separately. An asterisk (*) indicates a species with an associated narrative profile in the next chapter.

**Breeding Habitat Preferences** (in first parenthesis, but not included in Aquatic and Shoreline Species sublist)

DC = Disturbed habitats and human constructions

FE = Forest edge and shrubby habitats

G = Grassland habitats (primarily tallgrass prairie)

RIP = Riparian (woody or herbaceous) shoreline habitats

UGF = Upland and gallery forest habitats

X = Specific nest substrates (such as rock, gravel, sand, clay)

*Fig. 9. Upland sandpiper and prairie phlox*

**Seasonal Status Categories** (in second parenthesis, or the only parenthesis in Aquatic and Shoreline Species sublist)

M = Spring and fall migrant

SM = Spring migrant, fall migration data lacking or inadequate

WM = Wintering migrant

PR = Permanent resident (reported every month)

SR = Summer resident (at least into June), presumed or known breeder

V = Vagrant (reported during one single-week period only, including all eBird years to 2018)

**Relative Occurrence Frequencies**
The percentages included in the second or only parenthesis indicate the proportion of weeks over a hypothetical 48-week year (assuming four weeks for each of 12 months) during which a species has been reported at Spring Creek Prairie on eBird's weekly observation summary (as of 2018). This percentage provides a minimum estimate of the probability of encountering a particular bird (from 4 to 100 percent) when visiting Spring Creek weekly over a 12-month period. For migrant species the probabilities for sightings would be variably higher if visits occurred only during the species' reported periods of seasonal presence.

**Terrestrial Species**

PHASIANIDAE: PHEASANTS, GROUSE, AND TURKEYS

**Ring-necked pheasant,** *Phasianus colchicus* (G) (PR; 79%)*
**Greater prairie-chicken,** *Tympanuchus cupido* (G) (PR; 58%)
**Wild turkey,** *Meleagris gallopavo* (FE) (PR; 40%)

ODONTOPHORIDAE: NEW WORLD QUAIL

**Northern bobwhite,** *Colinus virginianus* (G) (PR; 79%)

CATHARTIDAE: AMERICAN VULTURES

Turkey vulture, *Cathartes aura* (G) (SR, Mar–Oct; 54%)*

ACCIPITRIDAE: KITES, HAWKS, AND EAGLES

Osprey, *Pandion haliaetus* (RIP) (M, Apr–May, Sep–Oct; 8%)
Bald eagle, *Haliaeetus leucocephalus* (RIP) (M, Dec–Mar; 23%)
*Northern harrier, Circus cyaneus* (G) (WR, Sep–May; 56%)
Sharp-shinned hawk, *Accipiter striatus* (UGF) (V)
**Cooper's hawk,** *Accipiter cooperii* (UGF) (SR, Mar–Nov; 44%)*
Broad-winged hawk, *Buteo platypterus* (UGF) (V)
*Swainson's hawk, Buteo swainsoni* (G) (M, Apr–May, Oct; 15%)
**Red-tailed hawk,** *Buteo jamaicensis* (UGF) (PR; 90%)*
*Ferruginous hawk, Buteo regalis* (G) (V)
Rough-legged hawk, *Buteo lagopus* (G) (WM, Oct–Feb; 29%)
Golden eagle, *Aquila chrysaetos* (G) (V)

FALCONIDAE: FALCONS

American kestrel, *Falco sparverius* (G) (PR; 58%)
Merlin, *Falco mexicanus* (G) (WR, Oct–Feb; 15%)
Peregrine falcon, *Falco peregrinus* (G) (M, May, Sep; 4%)

GRUIDAE: CRANES

Sandhill crane, *Grus canadensis* (G) (V)

### CHARADRIIDAE: PLOVERS

**Killdeer,** *Charadrius vociferus* (G) (SR, Feb–Nov; 75%)*

### SCOLOPACIDAE: SANDPIPERS, SNIPES, AND PHALAROPES

**Upland sandpiper,** *Bartramia longicauda* (G) (SR, Apr–Oct; 29%)
*Long-billed curlew, Numenius americana* (G) (V)

### COLUMBIDAE: PIGEONS AND DOVES

**Eurasian collared-dove,** *Streptopelia decaocto* (DC) (PR; 83%)
**Rock pigeon,** *Columba livia* (DC) (PR; 27%)
**Mourning dove,** *Zenaida macroura* (FE, G) (SR, Feb–Nov; 69%)*

### CUCULIDAE: CUCKOOS AND ANIS

Black-billed cuckoo, *Coccyzus erythropthalmus* (UGF) (V)
**Yellow-billed cuckoo,** *Coccyzus americanus* (UGF) (SR, May–Sep; 25%)

### STRIGIDAE: TYPICAL OWLS

**Eastern screech-owl,** *Otus asio* (UGF) (V)
**Great horned owl,** *Bubo virginianus* (UGF) (PR; 33%)*
*Burrowing owl, Athene cunicularia* (G) (V)
**Barred owl,** *Strix varia* (UGF) (PR; 10%)

### CAPRIMULGIDAE: GOATSUCKERS

**Common nighthawk,** *Chordeiles minor* (G) (SR, May–Sep; 17%)
Whip-poor-will, *Caprimulgus vociferus* (UGF) (V)

### APODIDAE: SWIFTS

**Chimney swift,** *Chaetura pelagica* (DC) (SR, Apr–Oct; 27%)

### TROCHILIDAE: HUMMINGBIRDS

Ruby-throated hummingbird, *Archilochus colubris* (FE) (SR, Jun–Sep; 23%)

### FAMILY ALCEDINIDAE: KINGFISHERS

**Belted kingfisher,** *Megacycle alcyon* (RIP) (SR, Feb–Nov; 67%)*

### PICIDAE: WOODPECKERS

**Red-headed woodpecker,** *Melanerpes erythrocephalus* (UGF) (SR, Mar–Oct; 52%)
**Red-bellied woodpecker,** *Melanerpes carolinus* (UGF, FE) (PR; 90%)
Yellow-bellied sapsucker, *Sphyrapicus varius* (UGF, FE) (M, Mar, May, Sep–Oct; 10%)
**Downy woodpecker,** *Picoides pubescens* (UGF, FE) (PR; 96%)*
**Hairy woodpecker,** *Picoides villosus* (UGF, FE) (PR; 52%)
**Northern flicker,** *Colaptes auratus* (UGF, FE) (PR; 98%)*

### TYRANNIDAE: TYRANT FLYCATCHERS

Olive-sided flycatcher, *Contopus cooperi* (UGF, FE) (M, Aug–Sep; 4%)
**Eastern wood-pewee,** *Contopus virens* (UGF, FE) (SR, May–Sep; 40%)
Yellow-bellied flycatcher, *Empidonax flaviventris* (UGF, FE) (V)
Acadian flycatcher, *Empidonax virescens* (UGF, FE) (V)
Alder flycatcher, *Empidonax alnorum* (UGF, FE) (M, May–Jun; 6%)
**Willow flycatcher,** *Empidonax traillii* (UGF, FE) (SR; May–Sep; 31%)
Least flycatcher, *Empidonax minimus* (UGF, FE) (M, May, Jul)
**Eastern phoebe,** *Sayornis phoebe* (UGF, FE) (SR, Mar–Oct; 58%)
**Great crested flycatcher,** *Myiarchus crinitus* (UGF) (SR, Apr–Aug; 29%)

Scissor-tailed flycatcher, *Tyrannus for-ficatus* (G) (V)
**Western kingbird,** *Tyrannus verticalis* (FE) (SR, Apr–Aug; 29%)
**Eastern kingbird,** *Tyrannus tyrannus* (FE) (SR, Apr–Aug; 44%)

## LANIIDAE: SHRIKES

**Loggerhead shrike,** *Lanius ludovicia-nus* (G) (SR, May–Jul; 12%)
Northern shrike, *Lanius excubitor* (G) (WM, Nov–Mar; 37%)

## VIREONIDAE: VIREOS

**Bell's vireo,** *Vireo bellii* (FE) (SR, May–Sep; 39%)
Blue-headed vireo, *Vireo solitarius* (UGF) (M, May, Aug; 6%)
**Warbling vireo,** *Vireo gilvus* (UGF) (SR, May–Sep; 38%)
Philadelphia vireo, *Vireo philadelphicus* (RIP) (M, May, Sep–Oct; 10%)
**Red-eyed vireo,** *Vireo olivaceus* (UGF) (SR, May–Sep; 33%)

## CORVIDAE: JAYS, MAGPIES, AND CROWS

**Blue jay,** *Cyanocitta cristata* (UGF, FE) (PR; 75%)*
Black-billed magpie, *Pica hudsonia* (FE, G) (V)
**American crow,** *Corvus brachyrhyn-chos* (UGF, FE) (PR; 88%)

## ALAUDIDAE: LARKS

*Horned lark,* *Eremophila alpestris* (G) (SR, Mar–Oct; 31%)*

## HIRUNDINIDAE: SWALLOWS

**Purple martin,** *Progne subis* (DC) (SR, May–Sep; 12%)
**Tree swallow,** *Tachycineta bicolor* (FE) (SR, Apr–Oct; 44%)
**Northern rough-winged swallow,** *Stelgidopteryx serripennis* (G) (SR, Apr–Sep; 40%)

Bank swallow, *Riparia riparia* (X) (SR, Apr–Sep; 10%)
**Cliff swallow,** *Petrochelidon pyrrho-nota* (G) (SR, Apr–Sep; 33%)
**Barn swallow,** *Hirundo rustica* (DC) (SR, Apr–Oct; 56%)*

## PARIDAE: TITMICE

**Black-capped chickadee,** *Poecile atri-capillus* (UGF) (PR; 54%)*
Tufted titmouse, *Baeolophus bicolor* (UGF) (V)

## SITTIDAE: NUTHATCHES

Red-breasted nuthatch, *Sitta canaden-sis* (UGF) (V)
**White-breasted nuthatch,** *Sitta caro-linensis* (UGF) (PR; 96%)

## CERTHIIDAE: CREEPERS

Brown creeper, *Certhia americana* (UGF) (M, Mar, Oct; 4%)

## TROGLODYTIDAE: WRENS

Rock wren, *Salpinctes obsoletus* (X) (V)
**Carolina wren,** *Thryothorus ludovi-cianus* (UGF, FE) (SR; May–Nov.; 10%)
**House wren,** *Troglodytes aedon* (UGF) (SR, Apr–Oct; 52%)
Winter wren, *Troglodytes troglodytes* (UGF) (V)
**Sedge wren,** *Cistothorus platensis* (G) (SR, Apr–Oct; 42%)
Marsh wren, *Cistothorus palustris* (R) (SR, Apr–Oct; 29%)

## REGULIDAE: KINGLETS

Golden-crowned kinglet, *Regulus sa-trap* (UGF) (M, Jan, Mar; 4%)
Ruby-crowned kinglet, *Regulus calen-dula* (UGF) (M, Mar–May, Sep–Oct; 25%)

## SYLVIIDAE: GNATCATCHERS

Blue-gray gnatcatcher, *Polioptila caerulea* (UGF) (M, Apr–Jun, Aug; 27%)

## TURDIDAE: THRUSHES AND ALLIES

**Eastern bluebird,** *Sialia sialis* (FE, G) (PR; 92%)
Townsend's solitaire, *Myadestes townsendi* (UGF) (V)
Swainson's thrush, *Catharus ustulatus* (UGF) (M, May, Sep; 12%)
Hermit thrush, *Catharus guttatus* (UGF) (V)
Wood thrush, *Hylocichla mustelina* (UGF) (V)
**American robin,** *Turdus migratorius* (UGF, FE) (PR; 92%)

## MIMIDAE: MOCKINGBIRDS AND THRASHERS

**Gray catbird,** *Dumetella carolinensis* (FE) (SR, Apr–Oct; 50%)
**Northern mockingbird,** *Mimus polyglottos* (FE) (SR, Apr–Oct; 27%)
**Brown thrasher,** *Toxostoma rufum* (FE) (SR, Apr–Oct; 56%)

## STURNIDAE: STARLINGS

**European starling,** *Sturnus vulgaris* (FE) (PR; 92%)

## BOMBYCILLIDAE: WAXWINGS

Cedar waxwing, *Bombycilla cedrorum* (UGF, FE) (SR, Mar–Oct; 46%)

## CALCARIIDAE: LONGSPURS

Lapland longspur, *Calcarius lapponicus* (G) (V)
Smith's longspur, *Calcarius pictus* (G) (WM, Feb–Apr, Oct; 14%)
Snow bunting, *Plectrophenax nivalis* (G) (V)

## PASSERIDAE: OLD WORLD SPARROWS

**House sparrow,** *Passer domesticus* (FC) (PR; 46%)

## MOTACILLIDAE: PIPITS

American pipit, *Anthus rubescens* (G) (M, Apr–May, Oct; 8%)
*Sprague's pipit, Anthus spragueii* (G) (M, Apr, Sep–Oct; 15%)*

## FRINGILLIDAE: FINCHES

Purple finch, *Haemorhous purpureus* (UGF, FE) (WM, Nov–Jan; 6%)
**House finch,** *Haemorhous mexicanus* (FE, G) (PR; 92%)
Pine siskin, *Carduelis pinus* (FE) (M, Oct–Nov; 10%)
**American goldfinch,** *Carduelis tristis* (G) (PR; 100%)*

## PASSERELLIDAE: TOWHEES AND SPARROWS

Eastern towhee, *Pipilo erythropthalmus* (UGF, FE) (SR, Mar–Oct; 58%)
Spotted towhee, *Pipilo maculatus* (UGF, FE) (M, Jan–Jun, Aug–Dec; 55%)
American tree sparrow, *Spizelloides arborea* (FE, G) (WM, Oct–Apr; 46%)
Chipping sparrow, *Spizella passerina* (FE) (SR, Mar–Nov; 58%)
*Clay-colored sparrow, Spizella pallida* (G) (M, Apr–May, Aug–Oct; 19%)
***Field sparrow,*** *Spizella pusilla* (G) (SR, Mar–Nov; 54%)
*Vesper sparrow, Pooecetes gramineus* (G) (M, Apr–May, Sep–Oct; 17%)
***Lark sparrow,*** *Chondestes grammacus* (G) (SR, Apr–Oct; 31%)
*Savannah sparrow, Passerculus sandwichensis* (G) (M, Apr–May, Sep–Oct; 23%)
***Grasshopper sparrow,*** *Ammodramus savannarum* (G) (SR, Apr–Oct; 46%)*

Baird's sparrow, *Ammodramus bairdii* (G) (V)

**Henslow's sparrow,** *Ammodramus henslowii* (G) (SR, Apr–Aug; 35%)*

LeConte's sparrow, *Ammodramus leconteii* (G) (M, Aug–Oct; 13%)

Nelson's sparrow, *Centronyx nelsoni* (G) (V)

Fox sparrow, *Passerella iliaca* (FE) (M, Mar–Apr, Oct–Nov; 23%)

**Song sparrow,** *Melospiza melodia* (FE) (PR; 83%)

Lincoln's sparrow, *Melospiza lincolnii* (FE) (M, Feb–May, Sep–Oct; 32%)

Swamp sparrow, *Melospiza georgiana* (G) (M, Mar–May, Sep–Oct, Dec; 27%)

White-throated sparrow, *Zonotrichia albicollis* (FE) (M, Jan–May, Sep–Oct; 25%)

Harris's sparrow, *Zonotrichia querula* (FE) (WM, Sep–May; 58%)

White-crowned sparrow, *Zonotrichia leucophrys* (FE) (M, Mar–May, Sep–Oct; 25%)

Dark-eyed junco, *Junco hyemalis* (G) (WM, Oct–Apr; 56%)

ICTERIIDAE: CHATS

Yellow-breasted chat, *Icteria virens* (FE) (V)

ICTERIDAE: BLACKBIRDS, ORIOLES, AND ALLIES

**Bobolink,** *Dolichonyx oryzivorus* (G) (SR, May–Sep; 33%)*

**Red-winged blackbird,** *Agelaius phoeniceus* (G) (PR; 83%)*

**Eastern meadowlark,** *Sturnella magna* (G) (SR, Feb–Oct, Dec, Jan; 73%)

**Western meadowlark,** *Sturnella neglecta* (G) (PR; 81%)*

Yellow-headed blackbird, *Xanthocephalus xanthocephalus* (RIP) (V)

Brewer's blackbird, *Euphagus cyanocephalus* (FE, G) (M, Mar, May, Sep; 6%)

**Common grackle,** *Quiscalus quiscula* (FE, G) (SR, Mar–Nov; 54%)

Great-tailed grackle, *Quiscalus mexicanus* (FE, G) (M, Mar–Apr; 4%)

**Brown-headed cowbird,** *Molothrus ater* (FE, G) (SR, Feb–Oct; 62%)*

**Orchard oriole,** *Icterus spurius* (UGF) (SR, Apr–Aug; 35%)

**Baltimore oriole,** *Icterus galbula* (UGF) (SR, Apr–Sep; 42%)

PARULIDAE: WOOD WARBLERS

Ovenbird, *Seiurus aurocapillus* (UGF) (M, May; 4%)

Louisiana waterthrush, *Seiurus motacilla* (UGF) (M, May; 4%)

Northern waterthrush, *Seiurus noveboracensis* (UGF), (M, Apr–May; 4%)

Black-and-white warbler, *Mniotilta varia* (UGF, FE) (M, May; 6%)

Prothonotary warbler, *Protonotaria citrea* (R) (V)

Tennessee warbler, *Oreothlypis peregrina* (UGF, FE) (M, Apr–May, Sep–Oct; 12%)

Orange-crowned warbler, *Oreothlypis celata* (UGF) (M, Apr–May, Sep–Oct; 27%)

Nashville warbler, *Oreothlypis ruficapilla* (UGF, FE) (M, May, Sep–Oct; 15%)

**Common yellowthroat,** *Geothlypis trichas* (G) (SR, Apr–Oct; 48%)

American redstart, *Setophaga ruticilla* (UGF, FE) (M, May, Jul; 10%)

Magnolia warbler, *Setophaga magnolia* (UGF, FE) (M, May; 4%)

Bay-breasted warbler, *Setophaga castanea* (UGF, FE) (V)

**Yellow warbler,** *Setophaga petechia* (UGF, FE) (SR, Apr–Sep; 31%)*

Chestnut-sided warbler, *Setophaga pensylvanica* (UGF, FE) (V)

Blackpoll warbler, *Setophaga striata* (UGF, FE) (M, May; 4%)

Palm warbler, *Setophaga palmarum* (UGF, FE) (M, May; 4%)

Yellow-rumped warbler, *Setophaga coronata* (UGF) (M, Feb–May, Sep–Oct; 29%)

Wilson's warbler, *Cardelina pusilla* (UGF, FE) (M, May, Aug–Sep; 8%)

### THRAUPIDAE: TANAGERS, CARDINALS, AND GROSBEAKS

Summer tanager, *Piranga rubra* (UGF) (V)

**Northern cardinal,** *Cardinalis cardinalis* (FE) (PR; 94%)

**Rose-breasted grosbeak,** *Pheucticus ludovicianus* (UGF) (SR, Apr–Sep; 40%)

Blue grosbeak, *Passerina caerulea* (FE, G) (V)

Lazuli bunting, *Passerina amoena* (FE) (V)

**Indigo bunting,** *Passerina cyanea* (UGF) (SR, Apr–Aug; 33%)

*Dickcissel,* *Spiza americana* (G) (SR, May–Oct; 44%)*

## Aquatic and Shoreline Species

### ANATIDAE: WATERFOWL

Greater white-fronted goose, *Anser albifrons* (M, Feb–Mar, Oct–Nov; 23%)

Snow goose, *Anser caerulescens* (M, Feb–Mar, Nov; 25%)

Ross's goose, *Anser rossii* (SM, Feb–Mar; 10%)

Cackling goose, *Branta hutchinsii* (M, Feb–Mar, Nov–Dec; 15%)

**Canada goose,** *Branta canadensis* (PR, 67%)

Trumpeter swan, *Cygnus buccinator* (V)

**Wood duck,** *Aix sponsa* (SR, Feb–Oct; 60%)

Blue-winged teal, *Spatula discors* (M, Mar–Jun, Aug–Oct; 35%)

Northern shoveler, *Spatula clypeata* (SM, Mar–May; 21%)

Gadwall, *Mareca strepera* (M, Feb–Apr, Oct; 19%)

American wigeon, *Mareca americana* (SM, Feb–Apr; 12%)

Mallard, *Anas platyrhynchos* (M, Feb–Jun, Aug–Nov; 44%)

Northern pintail, *Anas acuta* (M, Feb–Mar; 6%)

Green-winged teal, *Anas crecca* (M, Feb–Apr, Sep–Oct; 23%)

Canvasback, *Aythya valisineria* (SM, Mar–Apr; 4%)

Redhead, *Aythya americana* (SM, Mar; 6%)

Ring-necked duck, *Aythya collaris* (M, Mar–Apr, Dec; 17%)

Greater scaup, *Aythya marila* (V)

Lesser scaup, *Aythya affinis* (SM, Mar–May; 15%)

Surf scoter, *Melanitta perspicillata* (V)

Bufflehead, *Bucephala albeola* (SM, Mar–Apr; 8%)

Common goldeneye, *Bucephala clangula* (V)

Hooded merganser, *Lophodytes cucullatus* (RSM, Mar; 4%)

Common merganser, *Mergus merganser* (RSM, Apr, Oct; 6%)

### PODICIPEDIDAE: GREBES

Pied-billed grebe, *Podilymbus podiceps* (M, Mar–Jun, Aug–Oct; 44%)

### RALLIDAE: RAILS AND COOTS

Virginia rail, *Rallus limicola* (V)

Sora, *Porzana carolina* (M, May, Sep–Oct; 10%)

American coot, *Fulica americana* (M, Feb–May, Oct; 23%)

### RECURVIROSTRIDAE: AVOCETS

American avocet, *Recurvirostra americana* (SM, Apr; 4%)

### CHARADRIIDAE: PLOVERS

Semipalmated plover, *Charadrius semipalmatus* (RSM, May; 4%)

## SCOLOPACIDAE: SANDPIPERS, SNIPES, AND PHALAROPES

Hudsonian godwit, *Limosa haemastica* (RSM, Apr; 4%)

Stilt sandpiper, *Calidris himantopus* (RSM, May; 4%)

Sanderling, *Calidris alba* (V)

Dunlin, *Calidris alpina* (V)

Baird's sandpiper, *Calidris bairdii* (V)

Least sandpiper, *Calidris minutilla* (M, Apr–May, Jul–Sep; 15%)

White-rumped sandpiper, *Calidris fuscicollis* (SM, May; 4%)

Buff-breasted sandpiper, *Calidris subruficollis* (V)

Pectoral sandpiper, *Calidris melanotos* (M, Apr, Jul; 13%)

Semipalmated sandpiper, *Calidris pusilla* (SM, Apr–May; 6%)

Long-billed dowitcher, *Limnodromus scolopaceus* (V)

Solitary sandpiper, *Tringa solitaria* (M, Apr–May, Jul–Sep; 19%)

Lesser yellowlegs, *Tringa flavipes* (M, Mar–May, Jul, Oct; 19%)

Greater yellowlegs, *Tringa melanoleuca* (M, Apr–Jun; 15%)

Willet, *Tringa semipalmata* (M, Apr, Jul; 4%)

Wilson's snipe, *Gallinago delicata* (M, Mar–May, Sep–Oct; 25%)

Spotted sandpiper, *Actitis macularia* (M, Apr–May, Jul–Sep; 27%)

American woodcock, *Scolopax minor* (M, Apr–May; 6%)

*Wilson's phalarope, Phalaropus tricolor* (SM, Apr–May; 8%)

## LARIDAE: GULLS AND TERNS

Franklin's gull, *Larus pipixcan* (M, Apr–May, Sep–Nov; 23%)

Ring-billed gull, *Larus delawarensis* (SM, Feb–Apr; 6%)

Black tern, *Chlidonias niger* (M, May–Jun; 4%)

Forster's tern, *Sterna forsteri* (M, May, Aug; 4%)

Least tern, *Sternula antillarum* (M, V)

## PHALACROCORACIDAE: CORMORANTS

Double-crested cormorant, *Phalacrocorax auritus* (M, Apr–May, Sep–Nov; 25%)

## PELECANIDAE: PELICANS

American white pelican, *Pelecanus erythrorhynchos* (M, Apr, Sep–Oct; 8%)

## ARDEIDAE: HERONS

American bittern, *Botaurus lentiginosus* (M, Apr, Oct; 4%)

**Great blue heron,** *Ardea herodias* (SR, Mar–Nov; 46%)

Great egret, *Ardea alba* (M, Apr, Jun, Sep; 13%)

Snowy egret, *Egretta thula* (V)

Little blue heron, *Egretta cerulea* (V)

Cattle egret, *Bubulcus ibis* (SM, May–Jun; 4%) (semiterrestrial)

**Green heron,** *Butorides virescens* (SR, Apr–Sep; 33%)

Black-crowned night-heron, *Nyctanassa nycticorax* (M, May, Sep; 4%)

## THRESKIORNITHIDAE: IBISES

White-faced ibis, *Plegadis chihi* (M, May, Sep; 4%)

# ⑥ Profiles of Selected Prairie Birds

## Northern Bobwhite

*Colinus virginianus*

*Identification:* The male bobwhite's familiar call, *bob-white!*, is probably known to nearly every rural Nebraskan. It can be heard along country roads from at least late April through June and provides for instant recognition. The pullet-sized calling male might be seen standing on a fencepost, his white throat and a white eye-stripe extending back from the lores to the nape contrasting with his otherwise mostly chestnut brown head and body. Females are much less conspicuous, with the white of the male's face and throat replaced by buff, and their entire body plumage an exercise in camouflage.

*Voice:* Besides the male's spring advertisement call, similar notes are used to reassemble scattered coveys. A wide variety of other softer, more "conversational" calls are also uttered by both sexes, mostly for maintaining contact with the mate, family, and, in fall and winter, the covey.

*Status:* To an alarming degree, the bobwhite's US population has declined greatly in the past several decades, most probably because of habitat lost and effects of pesticides on both the birds and their major food bases: seeds, leaves, and (especially for growing chicks) insects. Long-term Breeding Bird Surveys from 1966 to 2015 indicate that this species' population underwent an average annual decline of 3.48 percent, reflecting a long-term reduction of more than 90 percent over many parts of its range. Nebraska surveys indicate a 0.56 percent annual decline, based on a sample of 44 routes.

*Habitats and Ecology:* The bobwhite is not so much a grassland species as one that needs grassy cover for nesting, brush cover for escape, and a foraging source of native plants or cultivated crops, especially legume seeds, weedy herbs, and grains, plus a nearby source

*Fig. 10. Northern bobwhite*

of water. Bobwhites are strongly monogamous, and older birds reestablish pair-
bonds in early spring. They then seek out
a nesting site, and both sexes participate
in nest building. A scraped area is filled
with leafy materials, and grasses or other
leaves are arched over the nest to con-
ceal it from above. The female incubates
the clutch of about 14 eggs for 23 days.
However, the male takes over incubation
if his mate is killed, and both sexes normally participate in brood rearing. No
more than about five or six chicks are
likely to survive through the fledging period, and as the family matures they begin to merge with other families to form
fall coveys. Covey sizes of about 10 to 11
birds are an ideal size because that many
birds can form a tight roosting circle in
which the contact of their bodies might
help retain body heat and provide an entire 360-degree panorama view.

**Ring-necked Pheasant**
*Phasianus colchicus*
*Identification:* The common and familiar "ring-necked" male hardly needs description; females may be confused with
sharp-tailed grouse if their long tails and
more generally mottled brownish plumage are not noted. The powerful breast
and leg muscles of pheasants allow them
to run up to about 20 miles per hour, or
as fast as a human sprinter, and to flush
almost vertically into the air.

*Voice:* Males utter a distinctive crow-
ing call, a double-noted *caw-cawk*, during
late winter and spring that can be heard
for more than half a mile. Pheasant calls
are highly diverse and number an estimated 16 to 24 types, a remarkable count
for a nonpasserine bird lacking the complex vocal anatomy of songbirds.

*Status:* A common permanent resident but non-native species, the ring-
necked pheasant was introduced to the
United States from China in the late
1880s and early 1900s. Long-term Breeding Bird Surveys from 1966 to 2015 indicate that this species' population has
undergone an average annual decline
of 0.64 percent. Nebraska surveys indicate a 1.78 percent annual decline,
based on a sample of 51 routes. Mineau
and Whiteside (2013) concluded that
this species has undergone the fourth-

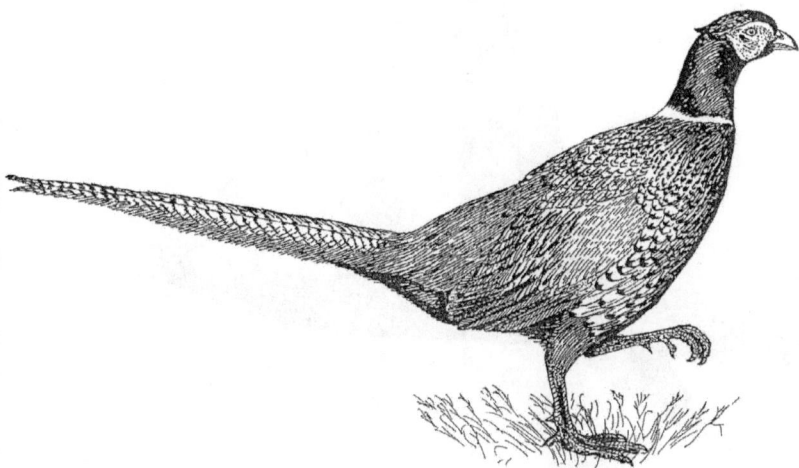

*Fig. 11. Ring-necked pheasant*

greatest long-term national population reduction among all ten grassland species that they studied and was declining in at least 19 states.

*Habitats and Ecology:* Breeding occurs mainly in native grasslands, along edges of woodlands and marshes, in irrigated agricultural areas, and among small patches with tall grass and weedy forbs. Nesting season in Nebraska extends from about mid-April to mid-June. Males are polygynous and attempt to attract several females into their harems. Although the females nest within their male's large home range, he does not participate in nest building or protection of the female, her nest, or the young. The clutch size varies from 7 to 15 eggs, and incubation begins when the clutch is complete. Brooding lasts 23 to 27 days, and the chicks are highly precocial at hatching. They fledge at 12 to 14 days and often begin to roost in trees when half-grown. Pheasants do not compete significantly with native grouse or quails. The recent declines of these native species are clearly the result of habitat losses and probably extensive pesticide use. It is fortunate that this single species of pheasant was so successful in adapting to grasslands in North America and Europe because most of the other nearly 50 pheasants of China and eastern Asia have become rare or endangered in their homelands (Johnsgard, 2017).

## Mourning Dove
*Zenaida macroura*
*Identification:* The mourning dove is one of Nebraska's commonest birds, and both its appearance and voice should be familiar to everyone. The recent incursion of the Eurasian collared-dove (*Streptopelia decaocto*) might cause confusion, but

*Fig. 12. Mourning dove*

this species has a more rounded rather than pointed tail and a distinctive black half-collar on the hind neck.

*Voice:* The mourning dove's courtship song is a four- or sometimes five-note *ooh, HOO, hoo, hoo*, whereas the Eurasian collared-dove's is a frequently repeated three-note *coo, coo!, cup.*

*Status:* In *The Second Nebraska Breeding Bird Atlas* (Mollhoff, 2016), the mourning dove ranked as the most frequently reported bird species, with a 99 percent occurrence rate in all the 557 blocks surveyed, and was judged to be the most numerous bird in the state. Long-term Breeding Bird Surveys from 1966 to 2015 indicate that this species' population underwent an average annual decline of 0.29 percent. Nebraska surveys indicate a 0.22 percent annual decline, based on a sample of 51 routes. Although morning doves are heavily hunted in the southern states, it seems unlikely that hunting has been a major factor in the very slow long-term decline of this species.

*Habitats and Ecology:* Mourning doves are notable for their high adaptability, which is the secret of their success. They are able to breed in all of Nebraska's terrestrial habitats, from the driest arid plains of the Panhandle to the wet riparian forests of the Missouri River valley, placing their nests on the ground, in bushes, or in trees. They also are persistent renesters; although they have clutches of only two eggs, they nest repeatedly from late spring to fall, and in Nebraska might manage to complete four breeding cycles in a single breeding season. Nesting in Nebraska extends from about mid-April to early September, a nearly five-month nesting span, so fitting in even four broods might be optimistic, inasmuch as about a month is required for each nesting cycle. As the breeding season ends, fall flocks begin to form, and a variably extensive migration southward occurs, depending on the severity of the weather. A few doves regularly overwinter in the Lincoln area.

## Killdeer
*Charadrius vociferus*

*Identification:* The most widespread and common of the North American plovers, the killdeer is easily recognized by its rusty brown tail and double breast band, together with its persistent calling, especially during the breeding season. Adults feign injury by performing "broken-wing" displays when their nest or young are threatened, effectively luring most intruders away.

*Voice:* An incessant *kill-dee* call is uttered mainly during the breeding season and often by the male as he makes high circling flights that might cover a half mile in diameter. Males also utter *di-yeet* or *di-yit* calls from conspicuous locations in their territory.

*Status:* The killdeer is a common summer resident, remaining well into fall and arriving early in spring. Long-term Breeding Bird Surveys from 1966 to 2015 indicate that this species' population underwent an average annual decline of 1.09 percent. Nebraska surveys indicate a 0.80 percent annual decline, based on a sample of 51 routes.

*Habitats and Ecology:* This species is widely distributed in open landscapes, including roadsides, reservoirs, ponds, gravel pits, golf courses, and suburban lawns. Gravelly areas with rocks about the size and color of the birds' eggs are favored locales. Nesting on rooftops sometimes occurs where gravelly habitats are absent. Killdeers forage visually on surface-dwelling insects, such as beetles, rather than probing for invisible foods in the manner of sandpipers and snipes. Nesting in Nebraska extends from about early May to late June, enough time for a single breeding effort. During pair formation, the male makes a series of shallow scrapes on the ground, the last of which is used by the female for depositing eggs. Shallow water is usually nearby, and stones that are 5 to 10 millimeters in diameter are selectively used for nest lining. Nearly all killdeer nests have clutches of four eggs, which are incubated by both sexes. The incubation period is 21.5 days, and initial flights occur about 30 days after hatching (Johnsgard, 1981). Killdeers remain into the fall and also are usually the first shorebird to arrive in southern Nebraska in spring.

**Great Blue Heron**
*Ardea herodias*
*Identification:* This heron is the largest of the common herons of the area and is mostly bluish gray with a black crown-stripe; a long, narrow crest; and a long, yellow bill. It flies ponderously with its long legs trailing and the head held back on the shoulders. During the summer it may be seen perching atop nesting trees; otherwise, it is usually found standing in shallow water, searching for aquatic prey.

*Voice:* This species is mostly silent except during the breeding season when it uses a *roh-roh-roh* call at the nest site and a croaking call when threatened by rival males or other intruders.

*Status:* A common summer resident, this heron sometimes remains into late fall. Long-term Breeding Bird Surveys from 1966 to 2015 indicate that this species' population underwent an average annual increase of 0.87 percent. Nebraska surveys indicate a 2.38 percent annual increase, based on a sample of 48 routes.

*Habitats and Ecology:* This species occurs along riparian areas where there are fish as well as suitable trees for constructing stick nests near their crowns. Large cottonwoods near water are a favored location for colonial nesting colonies, but sometimes the nests may be sit-

uated a mile or more from water. Nesting in Nebraska extends from May to late June. Older males return before the females and begin to refurbish old nests in preparation for their arrival. Pair-bonding involves a variety of complex postural displays, often performed mutually. After pairing has occurred, the male brings in more materials and the female finishes the nest. Typically four eggs are laid, and incubation by both sexes lasts 25 to 29 days. The young are fed and tended by both parents. The young initially begin to fledge at about 60 days and leave the nest at 64 to 90 days. Herons remain in southern Nebraska until fall freeze-up.

**Turkey Vulture**
*Cathartes aura*
*Identification:* Usually seen in flight, this species soars for long periods on wings that are slightly uptilted and two-toned, with black feathers in front and gray behind. Their primaries expose long finger-like tips while gliding or soaring; their flexibility is believed to be useful in reducing aerial drag. The unfeathered head is reddish in adults and appears small relative to the size of the wings and entirely blackish body.

*Voice:* Vultures cannot vocalize because they all lack syringes; the syrinx is the vocal structure responsible for avian calls and songs. However, vultures utter a low hissing sound when disturbed.

*Status:* The turkey vulture is a common summer resident that winters in the southern United States and Mexico; it arrives in Nebraska after most spring frosts and leaves before hard fall frosts. Long-term Breeding Bird Surveys from 1966 to 2015 indicate that this species' population underwent an average annual increase of 2.25 percent. Nebraska surveys indicate a statistically questionable 10.23 percent annual increase, based on a sample of 40 routes. Probably this apparent rapid increase is the result of turkey vultures adapting to "town life," by moving into villages, towns, and even large cities for nighttime roosting and by finding carrion among roadkills along nearby highways.

*Habitats and Ecology:* The turkey vulture is a scavenger species that consumes the carcasses of mostly larger animals, such as livestock and deer, which it finds both visually and by using its remarkable olfactory abilities. Vultures can often be seen soaring above on thermals, with few if any wingbeats needed, owing to their broad wingspan relative to a relatively light body mass, resulting in low wing-loading. Nesting often occurs in abandoned buildings such as those at old farmsteads. The nests are also sometimes located on or near the ground, in hollow logs, or on large snags. Typically two eggs are laid, and incubation is shared by both sexes. The eggs hatch in 37 to 41 days, and there is a long fledging period of about 11 weeks. The nesting and brooding season in Nebraska extends from about late May to late August. Fall migration often begins in late September.

## Cooper's Hawk
*Accipiter cooperii*
*Identification:* Like other bird-eating (accipiter) hawks, this species is characterized by a long, rounded tail and broad, rounded wings that adapt it for rapid and maneuverable flying in forests. It is notably larger (crow-sized) than the jay-sized sharp-shinned hawk. Adults are blue-gray above with reddish barring on the chest and red eyes. Immatures are more streaked with brown below and have yellowish eyes. The rapid flap-flap-glide flight behavior of accipiter hawks distinguishes them from the buteos, which tend to employ slow flapping and prolonged gliding. Most of the time Cooper's hawks perch nearly invisibly in heavy vegetation, waiting for potential prey to appear, and then fly off the perch in high-speed pursuit.

*Voice:* The Cooper's hawk is silent for much of the year, but during the breeding season both sexes utter a *cak-cak-cak* alarm call during courtship and in defense of the nest. All accipiter hawks have very similar *cak* calls.

*Status:* This hawk is an uncommon permanent resident. Long-term Breeding Bird Surveys from 1966 to 2015 indicate that this species' population underwent an average annual increase of 2.24 percent. Nebraska surveys indicate a statistically weak 14.58 percent annual increase, which was based on a sample of only 10 survey routes. However, Christmas Bird Counts also indicate a recent increase in Nebraska's Cooper's hawks, especially relative to sharp-shinned hawks, and a marked increase in city nesting.

*Habitats and Ecology:* The Cooper's hawk is associated with mature forests. Like other accipiter hawks, it is a highly effective predator of birds up to about the size of a quail, and it will sometimes attack prey as large as grouse and pheasants. Mammals and reptiles are less frequent prey but can include animals as large as skunks, opossums, rabbits, and hares (Johnsgard, 1990). Pair-bonds are maintained indefinitely, and the pair sometimes reuses its old nest, with the female adding fresh materials for the nest lining. Nests are usually placed in tall trees, up to 60 feet above the ground. Typically four eggs are laid on alternate days, and the female begins incubation with the laying of the third egg. Hatching occurs at 36 days. The female tends the young, and the male provides all the food for her and the nestlings. Fledging by the substantially smaller males occurs at about 30 days, and by the larger females at about 34 days. The young become independent by about two months.

**Red-tailed Hawk**

*Buteo jamaicensis*

*Identification:* The plumages of red-tailed hawks are extremely variable geographically, with many color forms ranging from light (leucistic) to very dark (melanistic). The locally breeding eastern race is a rich brown above and white below with a streaked belly or a variable belly-band of dark splotches. A small percentage of wintering birds are variably melanistic, sometimes appearing entirely black (the overwintering Harlan's hawk race) or very pale (the Great Plains Krider's race). The rusty upper side of the tail of adults is diagnostic, but immatures have brown-banded tails. All age groups have blackish leading edges on their underwings, extending from the "armpit" to the "wrist," which is the best identifying characteristic in flight.

*Voice:* This familiar hawk utters a prolonged *kee-eeee-rr* call, often heard in flight.

*Status:* The red-tailed hawk is a common summer resident and migrant. Many of them migrate to the southern United States for the winter, although some winter in the area, and others enter Nebraska from farther north to winter here, including Harlan's hawks. Numbers vary, with migration peaks in early April and October. Long-term Breeding Bird Surveys from 1966 to 2015 indicate that this species' population underwent an average annual increase of 1.42 percent. Nebraska surveys indicate a 3.58 percent annual increase, based on a sample of 50 routes. Judging from Nebraska Breeding Bird Atlas data, red-tails have moved west in recent decades, with about as many breeding records in the western half of the state now as in the eastern half.

*Habitats and Ecology:* This raptor is a common buteo hawk that occupies a broad range of habitats extending to open country. However, trees, especially large cottonwoods in shelterbelts and woodlands, are favored sites. Red-tailed hawks are extremely beneficial hawks and are the most common and widespread buteo in North America. Their foods are thus highly diverse, but in a survey of 11 published studies, mammals ranged from 37 to 88 percent of prey taken, averaging 68 percent; birds from 4 to 58 percent, averaging 17.5 percent; reptiles (mostly snakes) and amphibians from 0 to 41 percent, averaging 7 percent; and invertebrates 0 to 21 percent, averaging 3.2 percent. When food biomass is taken into consideration, the primary role of mammals in the overall diet becomes much more evident, with rabbits and rodents being primary prey targets (Johnsgard, 1990).

**Great Horned Owl**

*Bubo virginianus*

*Identification:* The largest of the "eared" owls of the region, this species has a wingspread of almost four feet. Females are significantly larger than males. Owls have notably large eyes with round pupils that open widely in the dark, and most species are specialized for night vision but lack color vision. Their heads can swivel 180 degrees, allowing their parabolic-like frontal facial disks to receive sound waves equally, allowing for stereophonic sound reception and thus the ability to judge exact prey locations.

*Voice:* The usual call is a low hoot, *who-who-who!, who-whoo.*

*Status:* The great horned owl is a common permanent resident. Long-term Breeding Bird Surveys from 1966 to 2015 indicate that this species' population underwent an average annual decline of 0.81 percent. Nebraska surveys indicate a 0.69 percent annual increase, based on a sample of 48 routes. Great horned owls are very uniformly distributed across the state, the mark of a species with a high degree of adaptability.

*Habitats and Ecology:* A powerful and adaptable owl, this species occurs everywhere from riparian woodlands to wooded suburbs. Nesting locations are thus highly variable but often are in abandoned bird or squirrel nests, in tree crotches, or rarely even on the ground. Great horned owls nest very early in the season; as a result the young are hatched at the time small rodent populations are increasing rapidly. In a counterpart role to the red-tailed hawk, the greater horned owl is the most widespread of North American owls, and very probably the most common. Its prey is likely highly variable but is strongly skewed toward mammals. Rabbits and hares are favorite foods in most areas. In a comparison of ten studies, lagomorphs (rabbit and hares) composed 54 to 70.5 percent of all foods taken by biomass, followed by larger rodents at 9 to 39 percent, mice and voles at 7 to 23 percent, ducks and gallinaceous birds at 4 to 8 percent, and passerine birds at 1.5 to 4 percent. Nesting in Nebraska begins as early as late January, and the usual clutch size is 2 to 3 eggs. Most of the incubation is by the female. Like other owls, incubation begins with the first egg laid, so the owlets hatch at several-day intervals following about 26 days of incubation. The older chicks are able to dominate the younger ones as to food access, so the number of young successfully raised depends on local relative prey abundance. The young depart the nest by about five weeks of age but are unable to fly well until they are about ten weeks old. They remain dependent on their parents for a considerable additional period, and a high juvenile mortality rate is typical (Johnsgard, 1988).

## Belted Kingfisher
*Megaceryle alcyon*
*Identification:* This large (13 inches) and conspicuous bird is easily identified by a bluish crested head, a wide bluish upper breast-band, and white underparts. Females have a second rufous band across the lower breast that is separated from the anterior band by white. Kingfishers are always found near water and often hover above its surface before plunging in to capture fish or, rarely, some other prey.

*Voice:* In flight, kingfishers utter a dry rattling call that resembles the sound of a fishing reel. It can often be heard even in winter where open water is available.

*Status:* The belted kingfisher is an uncommon summer resident that remains well into late fall until waters freeze over. Long-term Breeding Bird Surveys from 1966 to 2015 indicate that this species' population underwent an average annual decline of 1.37 percent. Nebraska surveys indicate a 0.77 percent annual decline, based on a sample of 29 routes. Their distribution in Nebraska is closely tied to river systems, especially rivers with steep banks.

*Habitats and Ecology:* This species is found near water that is rich in fish populations, usually where nearby roadcuts, eroded banks, gravel pits, or other steep earthen exposures provide opportunities for excavating earthen tunnel nests and where nearby tree branches provide convenient perching and observation sites. If fishing waters dry up, crayfish, frogs, toads, salamanders, snakes, insects, small mammals, or young birds might be eaten. The birds typically choose nesting sites along streams where riffles provide habitat for the small fish they prey upon. Nesting tunnels are dug into steep banks of clay or silt; they may be 10 to 15 feet long and take up to three weeks to excavate. Sometimes the male digs a second short tunnel to use for resting and sleeping. The usual clutch size is 6 to 8 eggs, and both sexes participate in incubation over the 23- to 24-day period. Fed by both parents, the young leave the nest at 30 to 35 days. If the first nesting is a failure, a second nesting is attempted, with the pair sometimes even excavating a new nest tunnel.

## Downy Woodpecker

*Picoides pubescens*

*Identification:* This species closely resembles the less common hairy woodpecker but is smaller (7 inches in length to the hairy's 9) and has a shorter, slenderer beak that is about half as long as the head (whereas the hairy's beak is more than half the length of the head). In both species the plumage is mostly black and white, with extensive white on the back and rump, and white spotting on the wings. Males of both species differ from females only in having a small red nape patch.

*Voice:* Woodpeckers don't sing, but most species have several loud calls. Common downy calls are a high whinny descending in pitch and a sharp *pik*. Downies often drum against wood or metal, the noise serving the same territorial-proclamation function as the songs of other birds.

*Status:* The downy woodpecker is a common permanent resident. Long-term Breeding Bird Surveys from 1966 to 2015 indicate this species' population underwent an average annual increase of 0.03 percent. Nebraska surveys indicate a 0.27 percent annual increase, based on a sample of 45 routes. The species is broadly distributed across the state but is rare in the Sandhills and the arid southwestern Panhandle grasslands.

*Habitats and Ecology:* A wide variety of wooded habitats are used by this species, but it has a preference for open deciduous riparian woodlands. Downy woodpeckers favor nesting in cottonwoods. Both males and females participate in excavating a nest cavity with a round entrance about an inch wide and a cavity 6 to 12 inches deep. Old downy woodpecker nest holes are valuable to a number of small cavity-nesting birds, including the house wren, black-capped chickadee, eastern bluebird, tufted titmouse, and tree swallow, which cannot excavate their own nest holes. A clutch of 4 to 5 eggs is typical for downies. Both sexes incubate the eggs, the male often at night. The incubation period is 12 days, and the fledging period is 20 to 22 days. Ants, the larvae of beetles, and caterpillars are common summer foods. Seeds and fruits are probably important during winter.

## Northern Flicker

*Colaptes auratus*

*Identification:* This woodpecker is mostly barred brown with black-scalloped patterning, a black breast-band, and underparts spotted heavily with black. A white rump patch is visible in flight. The local eastern race exhibits a yellow tint on the tail and flight feathers, and males differ from females in having black "mustache" (malar) stripes.

*Voice:* Vocalizations include a *wick-a-wick-a-wick-a* call uttered during the breeding season and a *klee-yer* that can be heard year round.

*Status:* This woodpecker is a common permanent resident but is most evident during winter. Long-term Breeding Bird Surveys from 1966 to 2015 indicate that this species' population underwent an average annual decline of 1.33 percent. Nebraska surveys indicate a 2.01 percent annual decline, based on a sample of 49 routes.

*Habitats and Ecology:* Broadly distributed, flickers are unusual among local woodpeckers because much of their food, which consists of insects such as ants and beetles, is obtained by probing in the ground. Flickers are often found in cottonwood stands and riparian woodlands, especially where snags are available. There they excavate nests that later are often used by many other cavity-nesting songbirds, and by small tree-dwelling mammals such as squirrels and deer mice. There are typically 6 to 8 eggs in a clutch, and incubation requires 11 to 13 days. Fledging occurs at 25 to 28 days. Where both the red-shafted and yellow-shafted races occur together, as in central Nebraska, mixed matings are not rare, and intermediate plumage types as well as (probably) backcross plumages are common (Short, 1965). These intermediate genotypes evidently survive as well as "pures" and are fully fertile. The zone of overlap and introgressive hybridization in Nebraska is very wide, and birds showing one or more traits of red-shafted flickers often can be found in eastern Nebraska, just as yellow-shafted phenotypes often appear in western Nebraska.

## American Kestrel
*Falco sparverius*
*Identification:* This tiny (10-inch) falcon, once known as the "sparrow hawk" because of its small size, may commonly be observed perched on telephone wires. In flight it often hovers directly above possible prey, a trait it shares with rough-legged hawks. Males have a rusty back and tail, a contrasting black "mustache" on their white face, and bluish gray wings. Females are more brownish overall and have an earth-brown barred tail.

*Voice:* A shrill disturbance call—*killy, kitty,* or *kleee*—is repeatedly uttered by both sexes and is used under conditions of generalized excitement. The call is the likely basis for the name of Kitty Hawk, North Carolina, where these hawks are common hunters over the sand dunes, and where they perhaps even revealed aeronautic principles to the Wright brothers.

*Status:* The kestrel is an uncommon summer resident, and some birds may overwinter during mild years. Long-term Breeding Bird Surveys from 1966 to 2015 indicate that this species' population underwent an average annual decline of 1.39 percent. Nebraska surveys indicate a 2.01 percent annual increase, based on a sample of 48 routes. During the more recent *Breeding Bird Atlas* surveys, it was possible to confirm breeding records from nearly every Nebraska county.

*Habitats and Ecology:* This bird is an open-country falcon that nests in tree cavities previously excavated by woodpeckers or in natural cavities of large trees. It will also nest in rocky crevices or similar spaces. It is the only hawk that will nest in nest boxes or woodpecker holes. Its usual clutch size is 4 to 5 eggs laid at 2- to 3-day intervals. Incubation is performed by the female, with the male providing her and the nestling young with food. The eggs hatch in 29 to 30 days, and fledging occurs at about 30 days. The kestrel is the smallest and most insectivorous of Nebraska's falcons. However, on a biomass basis, birds are generally the most important single dietary component, with mammals second. Other prey types, such as reptiles and insects, are of less importance, although on a numerical basis insects and other invertebrates may constitute up to 99 percent of the entire prey base (Johnsgard, 1990).

## Horned Lark
*Eremophila alpestris*
*Identification:* This grassland-adapted songbird can often be found along sandy or dusty country roads, where its pale brown upperpart coloration matches a dead-grass background, making it unlikely to be seen until it takes flight, when it exposes a white-edged black tail. Its distinctive face pattern, with broad black malar "mustaches," a black breast-band, and short feathered "horns," is distinctive.

*Voice:* This lark's territorial song is uttered during extended flights (Fig. 7) or while perched on a fencepost and is a weak twittering that doesn't begin to match the famous flight song of its close European relative, the skylark (*Alauda arvenis*).

*Status:* This species has one of the broadest ranges of all North American songbirds, extending from high Arctic tundra to hot, barren deserts, and even ranging into Europe, where it is called the shore lark. In Nebraska it is most common in treeless areas of the Sandhills and Panhandle. Long-term Breeding Bird Surveys from 1966 to 2015 indicate that this species' population underwent an average annual decline of 2.46 percent. Nebraska surveys indicate a 0.47 percent annual decline, based on a sample of 51 routes. During the more recent *Breeding Bird Atlas* surveys, it was possible to confirm breeding records from nearly every Nebraska county. Mineau and Whiteside (2013) concluded that this species, along with the eastern meadowlark and grasshopper sparrow, has undergone the greatest long-term national population reduction among ten grassland species that they studied, declining in 25 states.

*Habitats and Ecology:* Horned lark nests are typically in a small hollow sheltered by a clump of grass or beside a rock. The female incubates the clutch of about four speckled eggs for 12 to 14 days, and the fledging period is about 9 to 12 days. After fledging, the young remain near their parents for a week or more (Johnsgard, 2001a). They are quite nondescript in appearance; at times I have puzzled over their identity until I was able to find their nearby parents. Horned larks are notably cold-tolerant, and they are often the only sign of life detectable for anyone driving Nebraska's snow-rimmed roads during the middle of winter.

## Sprague's Pipit
*Anthus spragueii*

*Identification:* This pipit is very hard to see among prairie grasses, especially dead grasses, because its upper plumage is pale, dead-grass brown, much like a vesper sparrow's. A narrow white eye ring and faintly striped breast, also somewhat resembling a vesper sparrow's, are among the few good fieldmarks. At Spring Creek Prairie, the species is present only briefly during spring and especially fall migration and is most likely to be seen along the higher hilltops between late September and the end of October.

*Voice:* The best advice for finding Sprague's pipit on its prairie breeding grounds in the northern Plains is to listen for the flight song, which consists of a series of musical *tzeee-a* notes, uttered continuously from a fairly high altitude as the bird circles widely over its territory, which is typically hilly mixed-grass prairie.

*Status:* This is one of the many grassland-dependent species suffering a long-term population decline. Long-term Breeding Bird Surveys from 1966 to 2015 indicate that this species' population underwent an average annual decline of 3.01 percent, reflecting a serious long-term population reduction. Nebraska is outside the species' breeding range, which extends south from the Canadian prairies to northern South Dakota.

*Habitats and Ecology:* As for the Baird's sparrow, mixed-grass prairie is the prime nesting habitat for this pipit. It constructs its highly concealed grass nest in a hollow within growing herbage, hidden from above with overhanging vegetation. The female incubates the clutch of 4 to 6 heavily spotted eggs for about 13 days, and another 12 to 14 days are needed for the young to fledge (Johnsgard, 2001a).

## Grasshopper Sparrow

*Ammodramus savannarum*

*Identification:* Grasshopper sparrows rarely fly any higher above the grass canopy than a low fencepost from which they sing their insect-like songs. The species' stout beak and low crown produces a rather flat-headed profile, and its lore (the area just above and in front of the eye) is yellowish.

*Voice:* Appropriately, the grasshopper sparrow produces a so-called "song" that might be easily mistaken for the mechanical stridulation sounds of a grasshopper. It is mostly a low, soft, monotone buzzing, which becomes inaudible at any great distance (see fig. 4). One might wonder if the sparrow is luring a grasshopper, if both are perhaps challenging one another, or if this similar sound is just a biological coincidence, which is the most likely explanation.

*Status:* Grasshopper sparrows are not limited to prairies but are most abundant in tall and mid-height grasslands, from the annual grasslands of California to the Atlantic coastal prairies. Long-term Breeding Bird Surveys from 1966 to 2015 indicate that this species' population underwent an average annual decline of 2.52 percent. Nebraska surveys indicate a 1.91 percent annual decline, based on a sample of 51 routes. During the more recent *Breeding Bird Atlas* surveys there were possible to confirmed breeding records from every Nebraska county. Mineau and Whiteside (2013) concluded that this species, along with the eastern meadowlark and horned lark, has undergone the greatest long-term national population reduction among the ten grassland species they studied, declining in 25 states.

*Habitats and Ecology:* Grasshopper sparrows not only sound like grasshoppers, they also feed to a large extent on grasshoppers. A study cited in Austin (1968) revealed that of 170 stomachs examined, grasshoppers formed 23 percent of the total food over eight months of the year, and 37 percent from May to August, with a peak of 60 percent in June. Nesting in Nebraska is extended, with eggs reported from about mid-May to mid-August (Mollhoff, 2016). Two broods are typical for the central Plains, with estimated nesting success rates (at least one chick fledged per nest) of 35 to 52 percent (Johnsgard, 2001a). Sixty-six of 190 (35 percent) grasshopper sparrow nests were found to be parasitized in *The Second Nebraska Breeding Bird Atlas* study (Mollhoff, 2016). Among five studies cited by Ortega (1998), the median parasitism rate was 11.8 percent, with the rates ranging from 0 to 50 percent.

## Henslow's Sparrow

*Ammodramus henslowii*

*Identification:* The elusive Henslow's sparrow is another grassland sparrow with a very weak song and inconspicuous plumage. It is similar to the closely related grasshopper sparrow, with a yellowish lore area, but it has a more heavily streaked breast pattern, an olive-green-tinted head with a black triangle above its ear-coverts, two thin "mustache" lines, and much richer brown hues on the back and wing-coverts.

*Voice:* Even quieter than the grasshopper sparrow, males of this species utter only a short and soft *se—lik* as a territorial advertising song (Fig. 6).

*Status:* In Nebraska the known nesting range of this sparrow is limited to the southeast and currently (2018) includes only four local sites, extending as far west as Mormon Island, east to northern Seward County and Spring Creek Prairie, and south to Pawnee County (Mollhoff, 2016). Long-term Breeding Bird Surveys from 1966 to 2015 indicate that this species' population underwent an average annual decline of 1.53 percent.

Breeding Bird Survey data for Nebraska are lacking.

*Habitats and Ecology:* This species' nest sites are usually among rank, often moist, areas of taller grass and taller weeds or shrubs. The nests typically have a base of dead grassy litter and are either hidden under a tuft of overhanging grass or attached to the stems of herbage up to 20 inches above ground. The female incubates a clutch of usually three to five brown-blotched eggs for 11 days, and both parents attend the nestlings during their 9- to 10-day fledging period. Double-brooding is typical. Brood parasitism by cowbirds is evidently rare. In a Missouri study the parasitism rate was only 5.3 percent, as compared to 8.8 percent for dickcissels and an overall rate of 8.1 percent for all passerine birds (Winter, 1999).

The Henslow's sparrow was first painted and named by J. J. Audubon in 1829, for a bird he had collected in Kentucky. John S. Henslow was a botany professor at Cambridge University and a mentor to Charles Darwin when he was a student there in the 1820s. Henslow visited with Audubon when he was in England during 1831, less than a year before Darwin left on his historic voyage on the *Beagle*. Darwin carried with him a copy of Charles Lyell's just-published first volume of *Principles of Geology*, a gift from the *Beagle*'s captain Robert Fitzroy, which Henslow urged Darwin to read carefully but not to blindly accept its conclusions. Lyell's book would have an enormous impact on Darwin's thinking about the age and history of the earth, and in turn would revolutionize biological principles. It was, in fact, a book that helped Darwin not only modernize biology but also influence world history.

## Dickcissel
*Spiza americana*
*Identification:* Males of this small sparrow-like songbird are instantly recog-

nizable by their meadowlark-like pattern of a yellow chest with a black bib. They can often be seen perched on the highest shrub or small tree within their territories. Females also have a yellow chest but lack the black bib. Both sexes additionally have distinctive chestnut-colored upper wing-coverts.

*Voice:* The territorial male dickcissel repeatedly "tells its name," uttering a sweet, lisping *dick, ciss ciss, ciss,* or variants thereof, depending on its local dialect (Fig. 6). It is one of the last spring migrants, usually arriving in late May from its northern South America wintering grounds. From then through June the prairies are alive with the songs of dickcissels.

*Status:* Long-term Breeding Bird Surveys from 1966 to 2015 indicate that this species' population underwent an average annual decline of 0.36 percent. Nebraska surveys indicate a 1.04 percent annual decline, based on a sample of 51 routes. During the more recent *Breeding Bird Atlas* surveys there were possible to confirmed breeding records from every Nebraska county except for Sioux County. The estimated North American dickcissel population during the 1990s was 32 million (Rich et al., 2004).

*Habitats and Ecology:* In spite of their abundance at Spring Creek, I have not yet encountered a nest there, although cowbirds seem to be able to find and parasitize them easily. Ortega (1998) reported dickcissel parasitism rates in 12 studies to vary from 7 to 100 percent of the nests, with a median rate of 52.9 percent. Among such data from a total of 135 host species tabulated by Oretega, only the Bell's vireo (*Vireo belli*), with a median rate of 50 percent in 15 studies, and the red-eyed vireo (*Vireo olivaceus*), with a median rate of 55 percent in 13 studies, had comparably high parasitism rates. Dickcissel nests may be on the ground or in a shrub or low tree as high as six feet off the ground; the latter site is probably much easier for cowbirds to find. The female alone builds the nest in about four days; male dickcissels are often polygynous and may be otherwise occupied. A clutch is typically of four eggs, which the female incubates for 11 to 12 days. Only the female feeds and tends the young, which leave the nest in 7 to 10 days, although they might not fledge until they are 11 or 12 days old. Double brooding is probably fairly common, which might help compensate for cowbird-caused or other nest failures (Johnsgard, 2001a). By July, when second nesting efforts might occur, cowbirds have typically stopped laying (after depositing some 40 to 50 eggs, the female cowbirds are probably exhausted). Dickcissel egg dates in Nebraska extend from mid-May to late July (Mollhoff, 2016). Sixty-two of 92 (67 percent) dickcissel nests were found to be parasitized in *The Second Nebraska Breeding Bird Atlas* study (Mollhoff, 2016). Among 31 studies cited by Ortega (1998), the median parasitism rate was 7.1 percent, with the rates ranging from 0 to 76.5 percent.

## Bobolink
*Dolichonyx oryzivorus*
*Identification:* Male bobolinks on the breeding grounds are easily identified by a black overall plumage with a contrasting white rump and upper wing-coverts and a pale yellow-buff nape. Females are a sparrow-like mixture of streaky browns above and yellowish buff below with faint breast streaks, a black crown with a central buffy stripe, and pink beak and foot coloration.

*Voice:* Territorial males utter a rollicking, bubbling *bob-o-link* song, usually during low flights over their grassy territory (Fig. 6). Other calls are produced throughout the year.

*Status:* Bobolinks are most common in wet meadows, a habitat type that is rare and vanishing rapidly in Nebraska because these highly productive habitats are being converted to crop production. Perhaps as a result, long-term Breeding Bird Surveys from 1966 to 2015 indicate that this species' population underwent an average annual decline of 2.06 percent. In contrast, Nebraska surveys indicate a surprising 1.80 percent annual increase, based on a marginally significant sample of 40 routes. During the two *Nebraska Breeding Bird Atlas* surveys, there was a concentration of breeding records in the eastern Sandhills. Bobolinks winter in central and southern South America, where they probably encounter high levels of pesticide use in agricultural areas while they forage; they are very fond of rice and no doubt suffer from agricultural pesticide spraying on both their wintering and breeding grounds.

*Habitats and Ecology:* Bobolinks breed in moist lowland meadows of mixed grasses and tall forbs, building their nests in shallow hollows that are effectively hidden by surrounding vegetation. The clutch of five to six greenish eggs, spotted with dark brown and a blackish tint, is incubated by the female for 11 to 13 days. The nestling period is 10 to 14 days, although fledging may not occur until a few days later (Baichich and Harrison, 1997). In Nebraska the nesting period (the duration of all nests with one or more eggs or nestlings present) extends from mid-May to mid-July (Mollhoff, 2016). In some areas second nestings, or at least renestings, are undertaken after the failure of an initial nest (Johnsgard, 2001a). Ninety-one of 192 (47 percent) bobolink nests were found to be parasitized in *The Second Nebraska Breeding Bird Atlas* study (Mollhoff, 2016). Among six studies cited by Ortega (1998) the parasitism rates ranged from 0 to 70 percent.

*Fig. 13. Western meadowlark*

## Western Meadowlark

*Sturnella neglecta*

*Identification:* Although it is easy enough to recognize a meadowlark, distinguishing the western meadowlark from the eastern species is much harder. Both sexes of the western meadowlark are generally paler than their eastern counterpart, especially on the flanks, where the brown streakings are smaller. The yellow of the throat region also extends farther up into the white lower cheek, producing a narrower white "mustache."

*Voice:* The differences in the territorial songs of the western and eastern meadowlarks are usually diagnostic. As noted earlier, the western's song is an extended, melodic sequence of rapid, fluty notes, uttered far too rapidly to try to put a series of words to it, whereas the eastern's song is much shorter and the notes by comparison sound slurred, expressed something like "spring-is-nearly-here." Westerns also have a rasping alarm note that they utter when flushed, which easterns lack.

*Status:* The state bird of Nebraska, the western meadowlark is still common in meadows and native prairies. During Breeding Bird Surveys in the early

2000s the species was judged to be the state's ninth most common breeding bird (Mollhoff, 2016). During those surveys 235 western meadowlark nests were reported, as compared to 99 of the eastern meadowlark. During the more recent *Nebraska Breeding Bird Atlas* surveys, there were breeding records from every Nebraska county. Long-term Breeding Bird Surveys from 1966 to 2015 indicate that this species' population underwent an average annual decline of 1.29 percent. Nebraska surveys indicate a 1.34 percent annual decline, based on a sample of 51 routes. The estimated North American western population during the 1990s was 32 million (Rich et al., 2004).

*Habitats and Ecology:* Mixed-grass prairies are prime habitat for western meadowlarks, but the birds do extend west into shorter grasses and sage-steppe, and east into wetter and taller prairies. Western meadowlark nests are well hidden by overhead grasses; in more than 70 years of bird-watching I have observed very few nests, although at various times flushed females have alerted me to the fact that a nest must be very near, and that I should take a detour. The clutch is usually five speckled white eggs, which the female incubates for 13 to 15 days. The nestling period lasts 11 or 12 days; double brooding is typical during the fairly long breeding period. Among 23 major North American host species reviewed by Johnsgard (1997), this species was most often parasitized by brown-headed cowbirds (47 percent of 294 nests). One hundred of 235 (43 percent) western meadowlark nests were found to be parasitized in *The Second Nebraska Breeding Bird Atlas* study (Mollhoff, 2016). Cowbird eggs are similar in shape and color pattern to those of meadowlarks but are smaller. Western meadowlark egg dates in Nebraska extend from early April to late July (Mollhoff, 2016).

## Red-winged Blackbird
*Agelaius phoeniceus*

*Identification:* Breeding-season male red-winged blackbirds are unmistakable with their red wing "epaulets" that contrast with an otherwise all-black plumage, providing a definitive fieldmark. Females are remarkably sparrow-like in appearance but are much more heavily streaked with dark brown underparts than any sparrow. Golden to rusty tints on the face, throat, and scapular feathers also help in the identification of females.

*Voice:* The male's territorial song, a sweet *gurg-a-leee*, uttered with wing- and tail-spreading, is distinctive. Like other icterids, various harsher calls are uttered throughout the year.

*Status:* This is one of the commonest of North American songbirds, with an estimated population of 210 million in the 1990s (Rich et al., 2004), making it the most abundant of the many North American icterid species. Long-term Breeding Bird Surveys from 1966 to 2015 indicate that this species' population underwent an average annual decline of 0.93 percent. Nebraska surveys indicate a 0.25 percent annual decline, based on a sample of 51 routes. During state breeding bird surveys in the early 2000s the species was judged to be the third most common breeding bird in Nebraska, and there were breeding records from every county (Mollhoff, 2016).

*Habitats and Ecology:* Although these birds will often nest in upland meadows, hayfields, and grasslands, their favorite breeding habitat consists of cattail-rich shallow marshes. Fragments of such marshes, such as wet and weed-choked roadside ditches, also provide an adequate substitute for breeding by a male and as many mates as he can attract. Females build sturdy nests around the stems of upright vegetation, such as cattails or rushes, using tightly woven leaves and sometimes mud for added support. Three to five pale blue eggs with scrawled darker markings are laid and incubated by the female for 10 to 12 days, followed by another 10- to 11-day nestling period prior to fledging. Two breeding cycles per season are typical, and a vigorous male may be able to support two or three mates simultaneously. Nesting in Nebraska extends from mid-May to mid-July (Mollhoff, 2016). This species is often parasitized by brown-headed cowbirds, with 47 percent of 45 nests parasitized in *The Second Nebraska Breeding Bird Atlas* study. Among 31 studies cited by Ortega (1998), the median parasitism rate was 7.1 percent, with the rates ranging from 0 to 76.5 percent.

**Brown-headed Cowbird**
*Molothrus ater*
*Identification:* Brown-headed cowbirds generally resemble other regional "blackbirds" but have shorter, blunter beaks and shorter tails. Males have a brown "hood" that extends down into the upper breast, and females are uniformly brown, both above and below. Even during the breeding season the birds are usually found in small, mixed-sex courting parties (no pair-bonds are formed) among which the males constantly engage in wing-spreading and bowing displays.

*Voice:* The courtship song of the male is a squeaky gurgle, accompanied by a bow. Other similar notes are made during the frequent courting chases.

*Fig. 14. Brown-headed cowbird female in a host species' nest*

*Status:* Long-term Breeding Bird Surveys from 1966 to 2015 indicate that this species' population underwent an average annual decline of 0.93 percent. Nebraska surveys indicate a 0.25 percent annual decline, based on a sample of 51 routes. The estimated North American cowbird population during the 1990s was 63 million (Rich et al., 2004). During breeding bird surveys in the early 2000s the species was judged to be the second most common breeding bird species in Nebraska, and there were breeding records from every county (Mollhoff, 2016).

*Habitats and Ecology:* This species is possibly the most destructive bird in North America, being responsible for the reproductive failures of millions of other songbirds annually as a result of its obligatory brood parasitism. Females lay their eggs almost indiscriminately in the nests of other birds, mostly grassland-nesting songbirds (Johnsgard, 1997), including all the grassland passerines that nest at Spring Creek Prairie. Cowbirds rely on the host species to accept and incubate the eggs and to rear their young, thereby condemning the host's young to probable death by starvation, unless they can recognize the alien eggs and expel or otherwise render them harmless by piercing them with their beaks. Cowbird egg shells are thicker than those of other nonparasitic icterids, making them more difficult for host species to pierce. Cowbird eggs also have an incubation period of 11 to 12 days, which is 1 to 3 days shorter than nearly all other North American icterids, and their slightly brown-spotted surface pattern closely resembles those of many vireos, warblers, and sparrows. Peer, Robinson, and Herkert (2000) found that western meadowlarks ejected 36 percent of inserted artificial cowbird eggs, and dickcissels ejected 11 percent. Ortega reported that the burying of cowbird eggs by a parasitized host is a defensive response most frequently observed among yellow warblers (in 36 percent of 678 nests) but also reported in 31 host species. Nest desertion by the parasitized species is also a common defensive response, especially among smaller host species unable to pierce the egg or expel it. Ortega (1998) reported nest desertion behavior in 19 frequently parasitized species, including dickcissels (16.7 percent of 18 nests), red-winged blackbirds (17 percent of 47 nests), grasshopper sparrows (22.2% of 9 nests), and eastern meadowlarks (46.4% of 28 nests). However, the actual reasons for all such desertions would be impossible to determine, including possible disturbance of the nest by other animals or researchers, and nest desertion rates of these same species with unparasitized nests averaged about 6 percent. During the second Nebraska Breeding Bird survey, 28 host species to cowbirds were found, the most important of which, in descending frequency of nests affected, were western meadowlark (87%), bobolink (86%), grasshopper sparrow (69%), dickcissel (60%), eastern meadowlark (42%), and red-winged blackbird (12%) (Mollhoff, 2016).

# 7 The Mammals of Spring Creek Prairie and Lancaster County

This list includes 43 mammal species reported (Jones, 1964, and later sources) from Lancaster County; 30 of these species have been reported from Spring Creek Prairie and are shown in **bold**. The taxonomic sequence of families is based on Genoways et al. (2008). Taxa within families are arranged alphabetically, initially by genus and secondarily by species. The list excludes historically extirpated species such as bison and elk. Measurements (given for only for shrews, bats, and some small rodents) refer to total body plus tail length or total wingspread (bats). An asterisk (*) indicates a species with an associated narrative profile in the next chapter.

DIDELPHIDAE: NEW WORLD OPOSSUMS

**Virginia opossum,** *Didelphis virginiana.* Common in woodland edges and towns.*

LEPORIDAE: HARES AND RABBITS

Black-tailed jackrabbit, *Lepus californicus.* Uncommon in drier grasslands.

White-tailed jackrabbit, *Lepus townsendii.* Extirpated from tall grasslands.

**Eastern cottontail,** *Sylvilagus floridanus.* Widespread, woods and tall grasslands.

*Fig. 15. Coyote*

## SORICIDAE: SHREWS

**Northern short-tailed shrew,** *Blarina brevicauda.* 3.7–5.5 inches. Common in grasslands.

Elliot's short-tailed shrew, *Blarina hylophaga.* 3.6–4.8 inches. Edge of range. Grasslands.

**Least shrew,** *Cryptotis parva.* 0.7–1.5 inches. Uncommon in grasslands, especially tall prairie.

Masked shrew, *Sorex cinereus.* 2.9–4.9 inches. Common in diverse habitats.

## VESPERTILIONIDAE: VESPER BATS

**Big brown bat,** *Eptesicus fuscus.* 3.4–5.4 inches. Widespread, diverse habitats; hibernator.

Eastern red bat, *Lasiurus borealis.* 3.7–5.0 inches. Widespread, diverse habitats; migratory.

Hoary bat, *Lasiurus cinereus.* 1.6–2.5 inches. Widespread, woodland edges; migratory.

Little brown bat, *Myotis lucifugus.* 2.3–4.0 inches. Eastern quarter of Nebraska, deciduous edges; hibernator.

Northern myotis, *Myotis septentrionalis.* 3.1–3.8 inches. Eastern half of Nebraska, wooded edges; hibernator.

**Evening bat,** *Nycticeius humeralis.* 3.3–3.8 inches. Eastern half of Nebraska, deciduous woods; migratory.

## FELIDAE: CATS

**Bobcat,** *Felis rufus.* Widespread but uncommon in woodlands.*

## CANIDAE: COYOTES, WOLVES, AND FOXES

**Coyote,** *Canis latrans.* Widespread in grasslands; declining.*

**Red fox,** *Vulpes vulpes.* Widespread; near woods, increasing in towns and cities.

## MUSTELIDAE: WEASELS, BADGERS, AND OTTERS

Long-tailed weasel, *Mustela frenata.* Widespread, grasslands and woods.

**Least weasel,** *Mustela nivalis.* Widespread, diverse habitats.

**Mink,** *Mustela vision.* Widespread, near rivers and wetlands.

**American badger,** *Taxidea taxus.* Widespread, grasslands, especially drier grasslands.*

## MEPHITIDAE: SKUNKS

**Striped skunk,** *Mephitis mephitis.* Widespread, diverse habitats, diurnal.

Spotted skunk, *Spilogale putorius.* Local, forest edges, nocturnal.

## PROCYONIDAE: RACCOONS

**Northern raccoon,** *Procyon lotor.* Common in riparian woods; also in suburbs and cities.*

## CERVIDAE: DEER, ELK, AND MOOSE

**White-tailed deer,** *Odocoileus virginianus.* Widespread, forest edges, grasslands.

## SCIURIDAE: SQUIRRELS AND MARMOTS

**Eastern fox squirrel,** *Sciurus niger.* Common in woodlands and towns.

Franklin's ground squirrel, *Spermophilus franklini.* Local in tall grasslands; hibernator.

**Thirteen-lined ground squirrel,** *Spermophilus tridecemlineatus.* Common in grasslands; hibernator.*

**Woodchuck,** *Marmota monax.* Common in woodland edges; hibernator.

### CASTORIDAE: BEAVERS

**Beaver,** *Castor canadensis.* Common in wooded wetlands and streams.

### GEOMYIDAE: POCKET GOPHERS

**Plains pocket gopher,** *Geomys bursarius.* Common in grasslands.

### HETEROMYIDAE: POCKET AND KANGAROO MICE

Hispid pocket mouse, *Chaetodipus hispidus.* 7.75–8.75 inches. Common in sandy grasslands.

### ZAPODIDAE: JUMPING MICE

**Meadow jumping mouse,** *Zapus hudsonicus.* 7.4–10.2 inches. Common in grasslands; hibernator.

### CRICETIDAE: NATIVE RATS AND MICE

**Prairie vole,** *Microtus ochrogaster.* 5.2–6.75 inches. Common in tall grasslands.

**Meadow vole,** *Microtus pennsylvanicus.* 5.5–7.5 inches. Common in moist grasslands.

**Muskrat,** *Ondatra zibethicus.* Common in wetlands and slow streams.

**White-footed mouse,** *Peromyscus leucopus.* 5.2–8.2 inches. Abundant in diverse habitats.*

**Deer mouse,** *Peromyscus maniculatus.* 4.6–8.75 inches. Abundant in all grassland habitats.*

**Western harvest mouse,** *Reithrodontomys megalotis.* 4.5–6.75 inches. Common in tall grasslands.

**Plains harvest mouse,** *Reithrodontomys montanus.* 4.25–5.6 inches. Common in dry grasslands.

Southern bog lemming, *Synaptomys cooperi.* 4.6–6.2 inches. Uncommon in wet meadows.

### MURIDAE: OLD WORLD RATS AND MICE

**House mouse,** *Mus musculus.* Introduced; common near humans.

**Norway rat,** *Rattus norvegicus.* Introduced; common near humans.

*Fig. 16. Bobcat*

# ❽ Profiles of Selected Prairie Mammals

## Virginia Opossum
*Didelphis virginiana*

*Identification:* This is North America's only marsupial, but the female's pouch is unlikely to be visible in the wild even with young inside. However, the long, naked tail is visible, as are the large black ears and long pink nose. Opossums are about the size of house cats, weighing from about 4 to 14 pounds. They are as likely to be seen in a tree as on the ground, and are active diurnally as well as nocturnally. They don't hibernate during winter but may retreat to a sheltered location during freezing weather. Nebraska is near the northern edge of their range, and the Virginia opossum is both the most northern opossum and the only member of the marsupial family to store fat.

*Voice:* Opossums hiss when threatened, and both sexes also utter clicking sounds in aggressive situations. Females also use clicking sounds when communicating to offspring. These sounds stimulate ambulatory babies to follow her, either by clinging to her back or belly, or by running along beside her.

*Status:* Opossums are widespread in Nebraska and seem to be one of the commonest of medium-sized mammals in cities, judging from the number of runover carcasses that can be seen along streets. They have poor emergency responses to oncoming vehicles; when dazzled by headlights they often react with a fatal "freeze" stance.

*Habitats and Ecology:* Opossums are nearly omnivorous; their foods range from carrion to live animals. They often eat grain, such as corn, in the winter and gradually shift to insects, other invertebrates, bird eggs, small mammals, and plant materials during summer. During fall and early winter they also consume fruits and berries. Females become sexually mature during their first year, and in the northern Plains typically have two litters per year, one in late January or February and another in May or June. The gestation period is about 13 days, after which 4 to 23 embryo-like young emerge in rapid succession, either singly or in groups. Their forelegs are developed well enough for them to climb up and into the female's marsupium, where they try to find one of the approximately 13 teats, although some of these are nonfunctional. The average number of pouched young is about 8.5. As the young grow, they are able to climb well, and their prehensile tail helps them maneuver in trees, but they can't hang upside-down from a branch by their tail alone (as charmingly portrayed in Pogo cartoons by Walt Kelly) for more than a few seconds, as I have personally determined. Opossums are short-lived, with few living more than two years. Their tendency to "play possum," by becoming immobile and feigning death when threatened, is of no survival value in the modern car-dominated world. Presidents and politicians might be able to survive indefinitely in the twenty-first century by dissembling but not opossums.

## Bobcat
*Felis rufus*

*Identification:* This widely distributed but highly elusive cat occurs throughout Nebraska. It favors habitats with good hiding places, such as rock caves, brush piles, fallen trees, and the like. It is easily distinguished from domestic cats by its bobbed tail (thus, bobcat), which is black-tipped. As in many other wild cats, the ears are slightly tufted and are contrastingly white behind. (Most cat species have similar white spots behind their ears; I have long wondered if they might serve as "guidelights" for kittens trying to follow behind their mother.) Bobcats weigh from 14 to 29 pounds, so they are larger than most domestic cats. Males average slightly larger than females.

*Voice:* Bobcats have vocalizations much like domestic cats, including yowling during the breeding season, a cough-bark when threatened, and a loud scream.

*Status:* Bobcats almost certainly occur in every Nebraska county and have been photographed at Spring Creek Prairie.

*Habitats and Ecology:* Bobcats are largely nocturnal, but I once saw one during the early morning hours near a Platte River crane roost. They are also solitary, forming only brief pair-bonds during the breeding season, which usually occurs during winter and spring. The gestation period is about 62 days, and the litter size ranges from 1 to 8 but averages 3. As in other cats, their eyes are closed at birth but open at about ten days, and by four weeks they are able to eat solid foods. They are weaned by 7 to 8 weeks and begin to follow their mother on short trips. By about seven months they start to disperse from their natal range and become fairly independent. Females become sexually mature at one or two years, but males are usually mature only at their second year. Home ranges of adults are highly variable in size, from 150 to nearly 5,000 acres. Their foods are also highly diverse, but rabbits and rat-sized rodents predominate, and prey ranges in size from mice to fawns. Fish, amphibians, reptiles, and birds are also eaten. Great horned owls have been reported to prey upon imma-

ture bobcats, while coyotes and mountain lions often kill adults.

## Coyote
*Canis latrans*
*Identification:* Coyotes are very dog-like, with a uniformly grizzled gray coat overall, except for a black-tipped tail and rusty to yellowish legs. The ears are prominent, and the tail is bushy and held almost between the legs when the animal is running. Coyotes typically weigh 20 to 35 pounds. (They rarely hybridize with large dogs to produce "coydogs.")

*Voice:* The communal howling calls of coyotes are familiar to farmers, ranchers, and outdoor-lovers. They are prolonged vocalizations, usually uttered during evening or early morning hours. The call consists of a few sharp barks, followed by a prolonged mournful howl, and ending with several short, sharp yips. Barking calls are used in threat situations. Other vocalizations and social posturing are essentially like those of domestic dogs. The coyote's Latin name means "barking dog." Its common name is from the native Mexican Nahuatl language, *coyotl*, thus the most authentic English pronunciation is "ky-oat" rather than "ky-o-tee."

*Status:* In spite of constant persecution, coyotes have managed to survive across most of their historic range and no doubt still occur in every Nebraska county. In some areas, mange has affected their heath and reduced their populations.

*Habitats and Ecology:* Coyotes are highly adaptable, occurring in habitats ranging from desert to woodlands, and even extending into tundra and tropical forests. They are mobile daytime predators, ranging long-term over distances up to 400 miles. They can trot at speeds of 25 to 30 miles per hour, run at speeds up to 40 miles per hour, and can leap as far as 14 feet. Although they are hated by most ranchers, they are highly beneficial to them because they primarily eat rodents. The livestock that are eaten are mostly scavenged or involve otherwise weakened animals. Coyotes are also valuable for their fur; in some years as many as 12,000 coyotes have been trapped in Nebraska to be sold as pelts. Coyotes form monogamous pair-bonds, and females come into estrus for only a few days once per year, between February and March, following a prolonged courtship period. There is a gestation period of 63 days, and an average litter size of six pups. The young are born blind and helpless but are weaned after being nursed for six or seven weeks, while the male brings food to the pair's den. The pair-bond lasts through the period of reproduction, and family ties thereafter are the basis for autumn and winter social hunting parties, as in wolf packs. Lifespans in the wild rarely exceed ten years; given Nebraska's gun-addicted culture and contempt for all predators, most Nebraska coyotes are probably very lucky if they reach five years of age.

## American Badger

*Taxidea taxus*

*Identification:* The badger is another of the Great Plains grassland mammals that is instantly recognizable, having a distinctively black-patterned face with a white line up the midline of the head and a broader buffy stripe extending from the mouth diagonally back below the eye to the ear. Badgers have long, grizzled, grayish brown dorsal fur and a low-slung body profile. Adults weigh 8 to 25 pounds, and males are larger than females. Badgers have powerful forelegs with partially webbed front toes and long claws, which provide for highly efficient digging.

*Voice:* Various hissing and grunting noises are uttered when a badger is threatened, but otherwise these are quite silent animals.

*Fig. 17. Badger*

*Status:* Badgers are widespread in Nebraska but are most common on the open plains where rodent populations are high. Badgers are sometimes shot for "sport," and at times bounties have been paid for them, but probably most are killed in traps set for other mammals. Their digging activities help with soil development, and the holes they produce are used for escape or as dens by many other animals. They are effective killers of venomous snakes and are relatively immune to rattlesnake venom.

*Habitats and Ecology:* Badgers range across many habitats and ecosystems but favor open areas of meadows, prairies, steppe grasslands, or other places where subterranean dens for breeding and semihibernation during winter can be dug. They are extremely adept at digging out ground squirrel and prairie dog tunnels; at times two badgers will collaborate, with one digging and the other waiting at another tunnel entrance to catch any escaping inhabitants. Badgers breed in the summer or fall, but the pair-bond lasts only until the female is fertilized. After conception she takes on all further responsibilities for reproduction, including protecting the offspring. The young are not born until March or April, owing to delayed implantation of the embryos. Two to four babies typically compose a litter; they are born helpless and blind but are fur covered at birth. The young are weaned when they are about two-thirds grown, and females may become sexually mature within a year of birth. Both sexes are highly mobile as adults, males having enormous home ranges of anywhere from 600 to 4,000 acres. Home ranges of males often overlap those of several females; anal scent glands are present that may help in social communication and coordinating sexual activities. Badgers can potentially live for up to 14 years, but there is

high mortality among the young, probably mainly as a result of starvation and human killing.

## Northern Raccoon
*Procyon lotor*
*Identification:* Another easily recognized mammal, raccoons have a white-bordered, black-masked face and a tail with 4 to 7 brown and black rings. They also have white-bordered black patches behind their ears but are otherwise rather uniformly brown, grayish, or reddish brown, varying somewhat with the season and the regional population. Adults weigh 12 to nearly 50 pounds. Northern populations are the largest and are able store body fat for winter survival.

*Voice:* Highly vocal, raccoons produce a very wide variety of sounds, including whimpering, purring, screaming, growling, hissing, snarling, and chuckling as well as a shrill, tremulous whistle that is used by adults during autumn.

*Status:* Raccoons are widespread, highly adaptable, and able to survive almost as well in cities as in the countryside. They are most common where water is present, such as at marshes and in riverine forests. Their foods are equally diverse, depending on their habitats. Raccoons vary from being carnivores to vegetarians, and consume such items as insects, fish, crustaceans, bird eggs, mollusks, earthworms, grain, fruit, and discarded kitchen wastes. Their front feet are relatively handlike and are adapted for holding and manipulating objects.

*Habitats and Ecology:* Raccoons in the northern Great Plains are probably most abundant in riverine woodlands, especially those with oaks, elms, and sycamores, where they can use tree hollows, squirrel nests, stumps, or rotten logs for dens. Individual raccoons might have several dens within their home range, which for males might exceed 4,800 acres and for females up to 2,400 acres. Mostly nocturnal, raccoons can move at speeds of up to 15 miles per hour, and long-term movements of more than 175 miles have been reported. Raccoons do not establish territories but often move in family groups of up to about six animals. Males become sexually active in January and February, moving from den to den in search of receptive females. Gestation requires 63 to 65 days, with most litters of three to four young being born in late April or early May, but rarely births do occur as late as October. Blind at birth, the eyes of newborns open at 18 to 29 days of age. The youngsters are weaned by eight to ten weeks and soon begin to follow their mother on foraging searches. They become sexually mature as yearlings. Average longevity in the wild is usually only a few years, although lifespans up to 12 years have been reported.

## Thirteen-lined Ground Squirrel

*Spermophilus tridecemlineatus*

*Identification:* This is the familiar "gopher" that is often seen at the edges of gravel roads, where its striped and spotted body is easily observed. Its upperparts consist of a series of about 13 alternating dark brownish and whitish stripes, the dark stripes marked with regularly spaced whitish spots. The legs are short and the ears are small, and the tail is fairly long but not bushy. Adults weigh about 4 to 10 ounces and are much heavier prior to fall hibernation than during spring.

*Voice:* When frightened, and as a warning call, this ground squirrel utters a loud bird-like trill or tremulous whistle.

*Status:* The most common ground squirrel in Nebraska, population densities in the northern Plains range from about 1 to 10 per acre, but higher num-

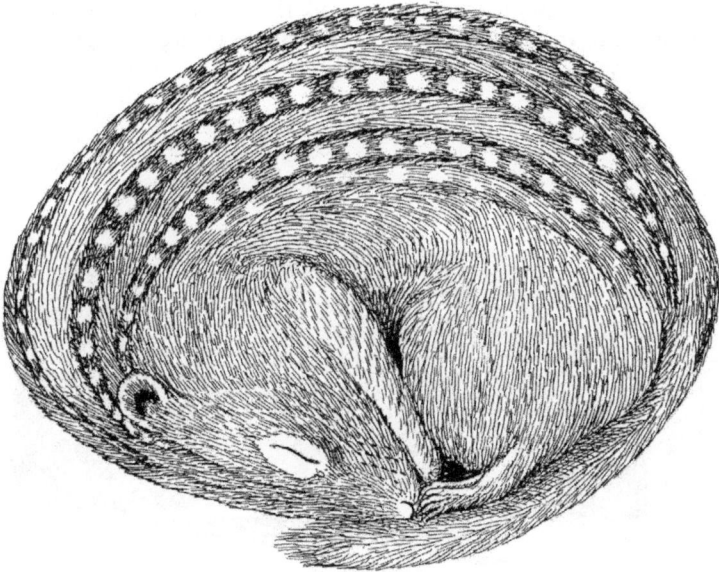

*Fig. 18. Thirteen-lined ground squirrel*

bers occur late in the breeding season as young emerge. In one Great Plains study the home range of males averaged 11 to 12 acres, and was largest during the breeding season, while those of females averaged 3 to 4 acres and was largest during pregnancy and lactation.

*Habitats and Ecology:* Much more associated with shortgrass and mixed-grass habitats than tallgrass prairie, this species is common statewide in grazed pastures, golf courses, and roadsides where the soil is well drained and can be easily excavated. Burrows may be up to 20 feet long and vary in depth from a few inches to as much as 4 feet. There are also short escape burrows scattered through the animal's home range. Except during spring, the entrances of burrows are not revealed by piles of soil. Above ground, there are grassy runways that are used to travel from one burrow to another, or for foraging. This species is among the least social of the ground squirrels, but adults have a greeting ceremony during which they touch noses, and scent markings are made by rubbing their faces over objects in their environment. Males emerge from their hibernation in late March or April, ready to mate with females when they emerge a week or more later. After mating, there is a gestation period of 27 or 28 days, and litters of 5 to 13 young are born in May or early June. In the northern Plains a single litter per year is produced. Hibernation begins in late September or early October.

**White-footed Mouse and Deer Mouse**
*Peromyscus leucopus* and *P. maniculatus*
*Identification:* White-footed mice and deer mice are so similar in appearance that an in-hand inspection is needed to determine the species. Both are variable geographically, by season and by age, in their upperpart colors, which vary from a gray to rich brown color above, but invariably they have sharply contrasting white underparts; this bicolored pattern even includes the long, sparsely haired tail. The ears are very large, and the eyes are similarly large and protruding. In

*Fig. 19. Deer mouse*

Nebraska the white-footed mouse can usually be separated from the deer mouse by the white-footed mouse's slightly larger mass (averaging about 30 g versus 22 g), and its longer tail (68–81 mm, ave. 75 mm vs. 43–77 mm, ave. 60 mm). The white-footed mouse's tail is longer than the length of the head and body and is less strongly bicolored than the deer mouse's.

*Voice:* These mice are not notably vocal (at least to the human ear), but young deer mice of two different subspecies were found to produce distress calls at frequencies of 3,600 to 26,500 Hz in a pulsed series of two to five notes lasting 0.1 to 0.2 seconds, the pulses separated by shorter pauses (Hart and King, 1966). It is also known that ultrasonic calls (35,000 Hz) are used by male deer mice prior to and after copulation (Pomerantz and Clemens, 1981). An alarmed white-footed mouse will rapidly drum its forefeet.

*Status:* Both of these species are extremely common in Nebraska; the deer mouse is the most widespread and probably most common North American rodent, and this is probably also true in Nebraska.

*Habitats and Ecology:* In the northern Plains, white-footed mice are almost exclusively found in wooded areas, such as riparian woods, shelterbelts, and fencerows, whereas the deer mouse is more adaptable and in the northern Plains occurs in open meadows, prairies, badlands, croplands, shelterbelts, hedgerows, and coniferous woodlands. Where white-footed mice are present in woodlands, deer mice are rare or absent; however, deer mice are found in woodlands that lack white-footed mice. Both species live in burrows. The deer mouse's burrow is often under rocks, among debris, or under fallen logs, where it makes grass- and vegetation-lined nests for sleeping, protection from cold, and rearing their young. Both species are mostly nocturnal. The home ranges of white-footed mice vary greatly but are rarely more than an acre, with females usually having smaller ranges than males. Deer mouse home ranges in the northern Plains vary widely, from 0.1 to 10 acres. Both species eat a variety of berries, seeds, fruits, nuts, and other vegetation as well as insects, other invertebrates, and occasional small vertebrates. Both species are active year round, but the period of reproduction is mostly during spring and fall. The gestation period of both species is 22 or 23 days, and litter sizes range up to nine but average about four. The babies' eyes open at about two weeks, and the youngsters are weaned at about four weeks. Both sexes are mature by about seven to eight weeks of age, so by fall several generations may be present in the population.

# ⑨ The Herpetiles of Spring Creek Prairie and Lancaster County

This list includes 24 species of reptiles and 10 species of amphibians (collectively informally called "herpetiles") that have been reported from Lancaster County. Sixteen of these species have also been reported from Spring Creek Prairie and are shown in **bold**. Taxonomic nomenclature is based on Fogell (2010) and the Center for North American Herpetology database. A field guide with current taxonomic treatment and excellent range maps was recently published by Robert Powell, Roger Conant, and Joseph Collins (2016). Fogell (2010) published a hard-copy identification guide for Nebraska species, with color photos, natural history information, and range maps, which can also be accessed online at http://snr.unl.edu/herpneb/. Ballinger, Lynch, and Smith (2010) published a comprehensive survey of the biology and distribution of Nebraska herpetiles of Nebraska. The number of specimen locations they reported for the species reported from Lancaster County is indicated in the list below by the abbreviation "BLS-*x*." Taxa are arranged alphabetically, initially by genus and secondly by species. Measurements refer to body plus tail length, except for turtles, which instead are measured for the length of the carapace (dorsal bony shell). Some alternate English and older Latin names are shown in parentheses. An asterisk (*) indicates a species with an associated narrative profile in the next two chapters.

## Amphibians
### (Toads, Frogs, and Salamanders)

### Order Anura: Frogs and Toads

Frogs and toads are easily recognized on the basis of their lack of a tail as adults ("Anura" means "without a tail"); moist, glandular skin; and four legs, the rear pair being much larger than the front

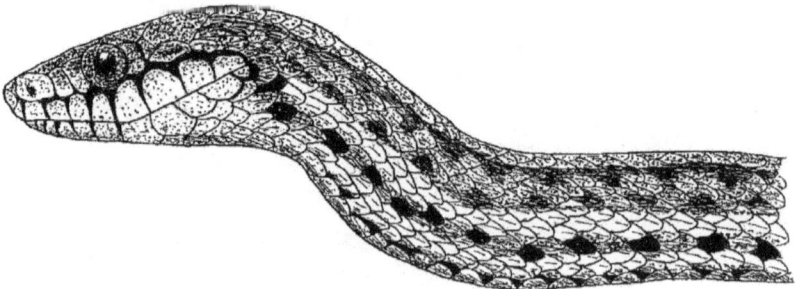

*Fig. 20. Plains gartersnake*

pair. Except for spadefoot frogs, all the frogs and toads of Nebraska have horizontal pupils. Like other amphibians, frogs and toads produce gelatinous eggs that develop in water, resulting in tailed, legless larvae with internal gills and mouthparts adapted for foraging on algae.

## BUFONIDAE: TRUE TOADS

Toads are semiterrestrial amphibians that have warty skin and a large parotoid gland on each side of the neck, either above the tympanum (ear drum) or on it, which secretes a highly poisonous mucus. The parotoid glands sometimes extend upward to the bony midpoint of the cranium (the cranial crest). Males vocalize in spring while inflating their throats, producing loud musical or metallic-sounding calls that serve to attract females.

**Woodhouse's (Rocky Mountain) toad,** *Anaxyrus (Bufo) woodhousei.* Widespread, common on dry grasslands. 2.5–4 inches. BLS-12*

Great Plains toad, *Anaxyrus (Bufo) cognatus.* Widespread, fairly common on dry grasslands. 3–4 inches. BLS-8*

## HYLIDAE: TREE FROGS

Tree frogs are small frogs that are often found in trees, being adapted for climbing by the presence of adhesive pads on their toes. One group of frogs (*Acris*) in this family doesn't climb trees; these species are variously known as chorus frogs, cricket frogs, and spring peepers. These species have reduced toe webbing and smaller toepads. Males of this group have a round, inflatable vocal sac that expands during calling, the local species producing buzzing (*Hyla*), cricket-like clicking (*Acris*), or sounds that resemble a comb's teeth being stroked (*Pseudac-*

*ris*). Some tree frogs, such as the common gray treefrog, are able to closely match their background by adjusting their skin color from gray, tan, or brown to bright green.

**Blanchard's (Northern) cricket frog,** *Acris (crepitans) blanchardi.* Widespread, common in wetlands. 0.6–1.4 inches. BLS-8

**Common (Cope's) gray treefrog,** *Hyla chrysocelis.* Widespread, in riparian woodlands. 1–2 inches. BLS-11*

**Boreal (Western) chorus frog,** *Pseudacris (triseriata) maculata.* Widespread, common in wetlands. 0.75–1.3 inches. BLS-12*

## RANIDAE: TRUE FROGS

Unlike tree frogs, true frogs lack toe pads and have a large rounded tympanum (external ear drum) behind each eye. The North American species all have long legs, pointed toes that lack terminal pads, and extensive toe webbing on muscular hind legs, which makes them powerful jumpers. They are also excellent swimmers. The adults are carnivorous, eating anything they can capture and swallow whole. Males typically call in chorus during spring, and during that season they have swollen forearms and thumbs that enable them to firmly grasp females during mating. Eggs are laid in water in long strings that might contain up to 20,000 eggs. From 6 to 24 months are needed for the hatched larvae to metamorphose and reach adulthood.

**American bullfrog,** *Lithobates (Rana) catesbiana.* Widespread in wetlands. 5–7 inches. BLS-9*

**Northern leopard frog,** *Lithobates (Rana) pipiens.* Widespread in wetlands, mostly north of the Platte River. 2.5–4 inches. BLS-10*

**Plains leopard frog,** *Lithobates* (*Rana*)
*blairi*. Widespread in wetlands. Visual recognition of *pipiens* from
*blairi* is difficult, but males differ
in calls (2–3-second snores by *pipiens*, 35–40-second chuckles by
*blairi*). 2.7–4 inches. BLS-12*

PELABATIDAE: SPADEFOOTS

Spadefoots are unique among Nebraska's
frogs and toads in having vertical pupils
as well as a black nail-like "spade" on
both of their hind feet. Spadefoots also
lack obvious parotoid glands, although
some people still have strong allergic re-
actions after handling them. Spadefoots
are found in sandy soils where they use
the "spades" on their rear legs to dig
into the sand and rapidly become hidden. Spadefoots can also inflate their
bodies when threatened, making them
more difficult to be swallowed by predators. Males produce rasping or snoring
calls in spring, usually while in temporary pools.

Plains spadefoot, *Spea bombifrons*.
Widespread, common on sandy
soils. 1.5–2 inches. BLS-4*

### Order Caudata: Salamanders

Salamanders are unique among amphibians in that they retain their long tails
into adulthood and develop four legs of
equal length. As aquatic larvae they also
have external gills, which in most species
are lost by adulthood, after which lung
breathing occurs. In a few permanently
aquatic species (and rarely in tiger salamanders) the external gills may be re-
tained throughout the animal's lifetime,
and their moist skin might also allow for
some oxygen exchange.

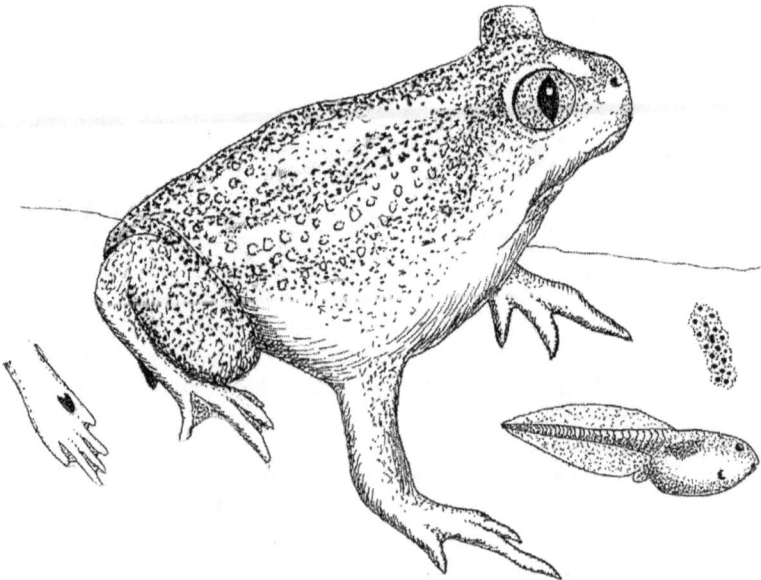

*Fig. 21. Plains spadefoot*

## AMBYSTOMATIDAE: MOLE SALAMANDERS

Mole salamanders (locally often called hellbenders) remain below ground for most of the year but move into pools and ponds for breeding. In Nebraska they sometimes can be found in water tanks used by livestock during their spring breeding period. The Nebraska population of the only regional species (*A. t. malvortium*) varies in skin color from brown to black, with bright yellow spots or bars on the sides; such barring might extend from the belly to the dorsal midline, producing a tiger-like pattern.

Barred tiger salamander, *Ambystoma (tigrinum) malvortium*. Widespread, common in wetlands. 6–8 inches. BLS-8*

## Reptiles (Turtles, Lizards, and Snakes)

### Order Chelonia: Turtles

All turtles are easily recognized by the presence of a bony dorsal "shell" (the carapace) and a corresponding but variably smaller ventral supporting bony structure (the plastron). In most turtles these protective structures are covered by thick, horny scutes. Turtles lack true teeth but have sharp rims along their upper jaws that serve to cut and tear their food. All turtles are egg-layers, typically depositing them in sandy soil, then abandoning them, requiring the hatchlings to independently dig their way out. All the North American turtles can defensively retract their fairly long necks back under the carapace.

## CHELYDRIDAE: SNAPPING TURTLES

Snapping turtles are notable for their hard, bony, and rough carapace, which is raised into three keel-like enlargements that extend from front to rear at its center, but the plastron is relatively small and cross-shaped. The tail is at least as long as the carapace and also has saw-toothed keels. Snapping turtles have large heads and powerful jaws. They can weigh up to 50 pounds, making them the largest (and much the most dangerous) of Nebraska's turtles. Like all turtles, they are oviparous, and females may lay clutches of up to more than 100 eggs, starting at about 10 to 12 years of age.

**Common snapping turtle,** *Chelydra serpentina*. Widespread, common in wetlands. 9–15 inches. BLS-1*

## EMBYDIDAE: POND AND BOX TURTLES

All of Nebraska turtles other than the snapping turtle and the two softshell turtles belong to the pond (or basking) and box turtle family, all of which have bony carapaces that are covered with smooth, horny scutes. Their plastrons are almost as large as their carapaces and are often distinctively colored or patterned. The tails are not as long as the carapaces. Two Nebraska species, the ornate box turtle and Blanding's turtle, have hinges on their plastrons that allow the front section to be elevated and the head to be more fully retracted when threatened. Most species are semiaquatic, and the aquatic species in this family tend to rest on floating logs or at the water's edge (thus the name "basking" turtles). The highly terrestrial box turtle rarely, if ever, enters water and is often found far from water. Some displaced box turtles have returned to home ranges from as far away as nine kilometers (Ballinger, Lynch, and Smith, 2010).

**Northern painted turtle,** *Chrysemys picta.* Widespread, abundant in wetlands. 5–7 inches. BLS-4*

False map turtle, *Graptemys pseudogeographica.* In the Missouri River and large tributaries; one Lancaster County record. 6–11 inches. BLS-1

Ornate box turtle, *Terrapene ornata.* Common on sandy soils. 5–6 inches. BLS-1*

### TRIONYCHIDAE: SOFTSHELL TURTLES

Softshell turtles are easily defined by their carapace being covered by leathery skin (sandpaper-like in one species, smooth in the other) rather than scutes; the carapace overlays the bony supporting skeleton. Softshell turtles also have a relatively long tail and large webbed hind legs that make them powerful swimmers. They have a flattened, pancake-shaped carapace, long neck, and a head with a long, tapering nose, which allows them to breathe by lifting just the tip above water. They eat crayfish, snails, insects, and other animal materials.

Smooth softshell, *Trionyx mutica.* Common in larger rivers. 6.5–14 inches. BLS-1

Spiny softshell, *Apalone spinifera.* Widespread in streams and wetlands. 8–20 inches. BLS-3

### *Order Lacertilia: Lizards*

### SCINCIDAE: SKINKS

Skinks are a family of lizards (body scaled, four legs, eyelids) that have small legs, smooth scales, and long, cylindrical bodies and tails. Many species have brightly colored tails that have fracture plates which allow the tail to break off when grasped, thereby permitting the animal to escape with minimal damage and to gradually regenerate its tail. All skinks are egg-layers, and females of the locally occurring species tend their clutches of up to about a dozen eggs during the 3- to 4-week incubation periods. The two skinks known from Lancaster County are diurnal, eat mostly insects and other arthropods, and are strongly patterned with multiple dark and white body stripes that extend into the tail.

Common five-lined skink, *Plestiodon fasciatus.* Local in moist woodlands. 4.5–7.5 inches. BLS-1

Northern prairie skink, *Plestiodon septentrionalis.* Local on grassy, rocky hillsides. 5–8 inches. BLS-1*

### *Order Serpentes: Snakes*

Snakes are in general easily distinguished from other reptiles because their scaled bodies and legless condition instantly separates them from all other reptiles except for a few legless lizards (glass lizards), which unlike snakes have movable eyelids and external ear openings. Snakes lack any auditory structures, so sounds can be perceived only from vibrations received through the body proper. Both jaws possess teeth, which are backward-slanted; in some species specialized teeth (fangs) exist that serve to inject venom into prey, which is then swallowed whole. Snake skull bones and jaws are connected only loosely so as to allow for the swallowing of very large prey. Most snakes have ventral plate-like scales that can be tilted and by means of muscular action and sinuous body movements used to facilitate locomotion.

## COLUBRIDAE: EGG-LAYING SNAKES

The most abundant and diverse group of Nebraska's (and the world's) snakes are the colubrid snakes, with about 2,700 species worldwide. These snakes are all nonvenomous, lack heat-sensitive (infrared-detecting) sensory pits, and have (mostly) round pupils. Although they have teeth in both jaws, the North American species lack specialized hollow fangs for injecting venom. Some species, however, do have grooved teeth connected to a poison-producing gland. None of the snakes in this family poses a serious threat to humans, although all of them eat whole animals ranging in size from insects to large rodents and other snakes. Most colubrids are egg-layers (oviparous) and produce clutches of up to about 25 eggs. Among regional species, only the water snake and crayfish snake are ovoviparous (give birth to internally hatched young).

**North American (Eastern) racer,** *Coluber constrictor*. On grasslands. 25–55 inches. BLS-5

Ring-necked snake, *Diadophis punctatus*. In wooded wetlands. 9–14 inches. BLS-1

Prairie kingsnake, *Lampropeltis calligaster*. On grasslands. 32–36 inches. BLS-1

Speckled (Common) kingsnake, *Lampropeltis (getula) holbrooki*. On wet prairies; edge of range. 36–48 inches. BLS-3

Western milk snake, *Lampropeltis triangulum*. On rocky grasslands. 24–36 inches. BLS-1

**Western fox snake,** *Mintonius vulpinus* (= *Elaphe vulpina*). Widespread, on prairies and woodlands. 36–50 inches. BLS-2

**Northern water snake,** *Nerodia sipedon*. Aquatic; common in wetlands. 36–44 inches. BLS-2

**Bullsnake (gophersnake),** *Pituophis catenifer*. Common on grasslands. 40–70 inches. BLS-7*

Graham's crayfish snake, *Regina grahamii*. In permanent wetlands. 25–32 inches. BLS-2

## NATRICIDAE: LIVE-BEARING SNAKES

Like the closely related colubrid snakes (and sometimes included as part of that family), members of the Natricidae assemblage have round pupils and lack facial sensory pits. However, they are livebearers, producing litters of rarely up to more than 80 offspring. Gartersnakes overwinter socially in collective underground retreats called hibernacula, which in northern latitudes might support hundreds of individuals. Mating may occur among the adults before they enter a hibernaculum or after they emerge in spring. Male gartersnakes typically emerge before the females and patrol the vicinity for weeks while waiting for the females to emerge.

**Dekay's brownsnake,** *Storeria dekayi*. Uncommon, in moist woodlands. 9–13 inches. BLS-4

**Plains gartersnake,** *Thamnophis radix*. Abundant in grasslands. 20–28 inches. BLS-9*

**Common gartersnake,** *Thamnophis sirtalis*. Abundant in grasslands. 22–32 inches. BLS-7*

Western ribbon snake, *Thamnophis proximus*. Semiaquatic; on wet meadows. 20–30 inches. BLS-1

**Lined snake,** *Tropicoclonion lineatum*. Prairies, in woodland edges; secretive. 8–12 inches. BLS-5

### VIPERIDAE: PIT VIPERS

All the members of this relatively dangerous snake family have large erectile and hollow upper teeth (fangs) that can inject venom, which is used to subdue their prey as well as defend themselves. Pit vipers have large triangular heads, eyes with vertical pupils, and paired infrared (heat-sensitive) pits located just behind the nostrils. All of these snakes specialize in eating warm-blooded vertebrates, although other snakes, lizards, frogs, and a variety of invertebrates may also be consumed. All of the most dangerous North American snakes belong to this family. A total of four pit vipers are found in Nebraska, with three of them confined to a few counties in southeastern Nebraska.

Massasauga, *Sistrurus catenatus*. A rattlesnake extirpated from Lancaster County; the last record was prior to 1942 (Hudson, 1942). 28–36 inches. BLS-4*

# ❿ Profiles of Selected Prairie Amphibians

**Woodhouse's Toad and Great Plains Toad**

*Anaxyrus woodhousei* and *A. cognatus*

*Identification:* Easily identified as toads by their stocky shape and highly warty skin, the Woodhouse's toad and Great Plains toad are very similar. Both occur in Lancaster County, but Woodhouse's (illustrated on the left in Fig. 22) is more likely to be found in urban yards and gardens. It has a pale line down the middle of the back, an unspotted abdomen, and its parotoid gland behind the eye extends farther forward, to the rear end of the bony cranial ridge that is present along the top of the head. The tympanum is inconspicuous but is about as large as the eye. When calling, the throat area (the vocal sac) is strongly inflated. Krupa (1994) has described the breeding biology of the Great Plains toad.

*Voice:* The Woodhouse's toad utters a nasal trill that is similar to the bleating of a calf or sheep and lasts 1 to 3 seconds, whereas the Great Plains toad produces a very loud, trilled scream that lasts up to 50 seconds and resembles the sound of riveting.

*Status:* The Woodhouse's toad is widespread and has been reported from all but four Nebraska counties.

*Habitats and Ecology:* Woodhouse's toad favors sandy areas but can be found anywhere that is fairly close to water. It is nocturnal, remaining in its burrow throughout the day. It emerges at night to feed on insects, especially beetles and ants, as well as snails, spiders, and other arthropods. Breeding begins in late spring, with male choruses lasting until midsummer. Eggs are laid in long strings of up to at least 25,000 eggs. They hatch in less than a week, and metamorphosis occurs 50 to 60 days after hatching.

**Common Gray Treefrog**

*Hyla chrysocelis*

*Identification:* This tiny tree-dwelling frog can change from bright green to brown within a few hours but often appears as a mottled gray or black dorsally. The inside of the thighs is bright yellow-orange.

*Voice:* In spring, males utter a high-pitched buzzy trill that lasts two to four seconds. A closely related and visu-

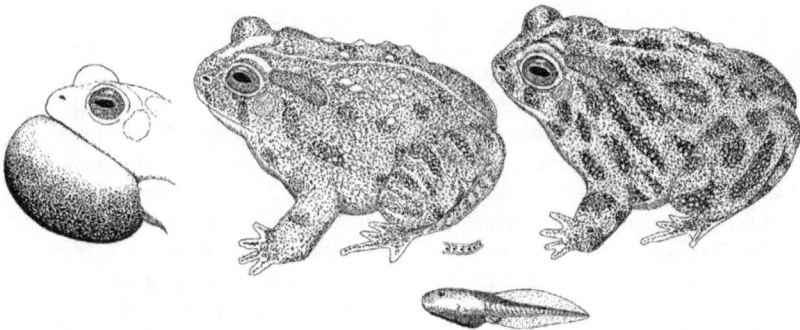

*Fig. 22. Woodhouse's toad (left) and Great Plains toad (right)*

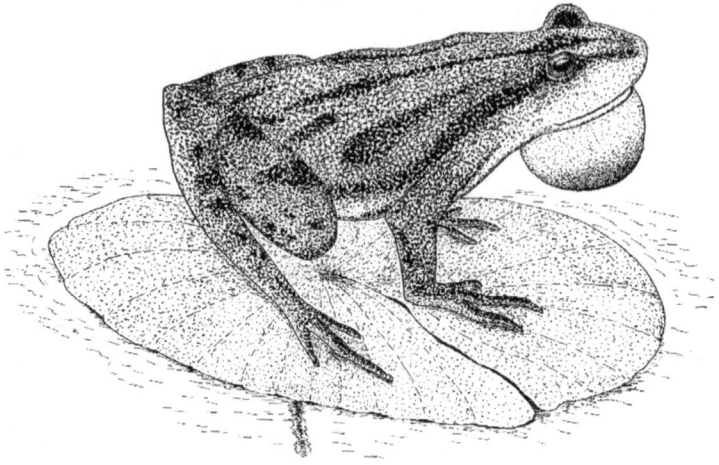

*Fig. 23. Boreal chorus frog*

ally identical form, *H. versicolor*, has a slower or lower-pitched voice and has been found in Iowa. It has not yet been reported from Nebraska (Ballinger, Lynch, and Smith, 2010), but it probably occurs here (Fogel, 2010).

*Status:* This frog ranges in southeastern Nebraska north to Dodge County and west to Lancaster and Jefferson Counties, with scattered occurrences west to Lincoln County and north to Knox County.

*Habitats and Ecology:* These treefrogs are active in Nebraska from about April through August, with breeding occurring from mid-April through June. They breed in permanent to temporary ponds near woods and those having weedy vegetation. Males call from the water's edge or surrounding vegetation and may perch up to ten feet above water. Females deposit up to 2,000 eggs in deep water, either singly or in clusters. By August the froglets have left the water and are absorbing what is left of their tails.

## Boreal Chorus Frog
*Pseudacris maculata*

*Identification:* This tiny frog barely exceeds an inch in length and is the only Nebraska frog having a white streak along the upper lip and three dark streaks along the back, plus wide stripes extending from the nose through the eyes and back along the flanks.

*Voice:* Males begin advertisement calling as soon as the snow melts, typically during early March in southern Nebraska. Their calls are uttered with the vocal sac greatly enlarged, while the frog is clinging to vegetation. Calling continues both throughout the night and during the day. The call is a mechanical trill that is similar to the sound produced by running fingers over the teeth of a comb. In Nebraska, singing may extend into August in reservoir-chilled water, but it normally ends by late April.

*Status:* This frog is widespread throughout Nebraska, having been reported from all but 15 counties. Water bodies of almost any size are used, including vernal ponds, flooded fields, sewage lagoons, and lakes. During the summer the frogs may also be found on grasslands and woods far from water.

*Habitats and Ecology:* Immediately after mating, females release their eggs, which are adhesive and cling to vegetation. Individual clutches may contain up to nearly 200 eggs, and the total seasonal ovarian production may be 500 to 800 eggs. Metamorphosis occurs after about six weeks—by June in Nebraska—and thereafter both juvenile and adult frogs feed on insects and other small arthropods. In a Colorado study, some frogs were found to survive for as long as six years.

## American Bullfrog

*Lithobates catesbiana*

*Identification:* The bullfrog is the largest frog in Nebraska, with adults at least five inches long. Adults vary in color from lime green to olive and have warty backs, often with reddish brown to blackish dorsal markings. The tympanum is large (especially in males) and is unmarked and conspicuous.

*Voice:* The breeding call of the male has been described as a low snore, sounding like *jug-o-rum*. It can be heard from May into July and serves both as a territorial and sexual advertisement signal.

*Status:* This frog has statewide distribution, having been recorded from all but about 26 counties. It is probably most abundant in eastern Nebraska, where there is more available surface water.

*Habitats and Ecology:* Bullfrogs usually can be found along the banks of almost any wetland, from ponds and marshes to slow rivers, where they patiently wait for prey to come within reach of their huge mouths. They try to capture anything they are able to swallow, from insects to other amphibians, turtles, small snakes, mammals, and birds. Breeding begins in late spring, when males advertise and defend their territories. After mating, females deposit clusters of 25,000 to 45,000 eggs as surface films on water. Following hatching, the larvae disperse and spend two years growing and completing their metamorphoses into adults, becoming torpid over the winters.

## Northern Leopard Frog and Plains Leopard Frog

*Lithobates (Rana) pipiens* and *L. blairi*

*Identification:* Recognizing leopard frogs is easy; they are the commonest amphibians in Nebraska. However, separating the Plains leopard frog from the northern leopard frog is another matter. Visual distinction between the two is difficult; both are smaller than bullfrogs, ranging in length from 2.5 to 4 inches, and their overall color is green to brown, with dark brown spots and blotches that are often edged with white. Two pale dorsolateral folds that extend from the back of the head are discontinuous toward the rear in the Plains leopard frog (upper right in Fig. 24), but these folds are continuous in the northern species. The Plains leopard frog also usually has a white spot in the middle of the tympanum, a color pattern lacking in the northern leopard frog, and the former species also has a generally more stocky body conformation.

*Voice:* Males of the two leopard frog species differ in their mating calls. In the Plains species the calls are a long (35–40 seconds) series of *chuck* notes that end in a longer *cu-u-u-uck*, the series sounding like a finger being rubbed over a balloon. In the northern species, the call is a long snore followed by two or three *chuck* notes, each series lasting only two or three seconds. Females of *L. pipiens* show differential responses

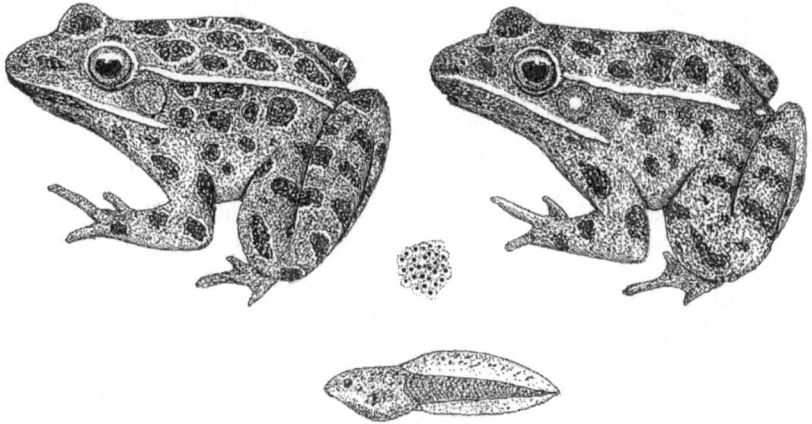

*Fig. 24. Northern leopard frog (left) and Plains leopard frog (right)*

to the calls of *L. blairi*, a presumed hybrid, and conspecifics (Kruse, 1981). Littlejohn and Oldham (1968) related the mating calls of the *Rana pipiens* complex to their taxonomy.

*Status:* Plains leopard frogs are common in the eastern two-thirds of Nebraska but are absent in the Panhandle and over most of the western counties north of the Platte River, where the northern species is common.

*Habitats and Ecology:* Plains leopard frogs begin their mating activities very early, usually by the end of March. Males call while floating at the water surface, beginning after sunset. The newly fertilized female deposits a globular cluster of 4,000 to 6,000 eggs that hatch within a few days. Metamorphosis into the adult frog occurs in 50 to 60 days. As adults the frogs eat a variety of insects and other invertebrates as well as vertebrates up to the maximum size that their mouths can accommodate. During summer the adults often wander about on land; prior to winter they bury themselves in mud and leaves at the bottom of ponds.

**Barred Tiger Salamander**
*Ambystoma malvortium*
*Identification:* This is the only salamander that occurs in Lancaster County and most of the rest of Nebraska, but the closely related smallmouth salamander (*A. texanum*) occurs in Otoe and Cass Counties, immediately to the east. That small and very rare salamander lacks the contrasting black and yellow spots and stripes of the tiger salamander. In parts of western Nebraska and the Sandhills, the tiger salamander's striped black color is reduced, and the resulting pattern is more one of dark blotches and spots over a yellow to greenish body.

*Voice:* Salamanders are mute.

*Status:* Tiger salamanders are common and widespread in the state, where they have been reported from all but 20 counties. However, they are nocturnal, and when living on land they emerge from their burrows to forage only at night. In western Nebraska they sometimes live in the burrows of prairie dogs. Sandy areas or loose soils with nearby water are favored habitats.

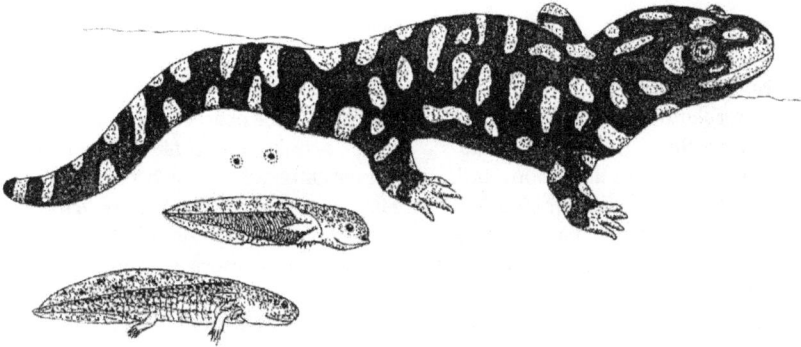

*Fig. 25. Barred tiger salamander*

*Habitats and Ecology:* Tiger salamanders begin breeding with the onset of early spring rains. At that time small ponds and stock tanks might become full of salamanders, but they avoid ponds with large fish; predatory fish often eat salamanders. Breeding occurs in early spring, usually during March and April in Nebraska. Mating occurs in water, and the eggs are also laid in water, either singly or in clusters. The larvae hatch within a few weeks and soon develop into voracious predators, even eating other salamander larvae. The larvae are distasteful to potential predatory fish, allowing the salamanders to survive in fish-rich waters. Most of them metamorphose on into the summer months, gradually losing their gills and tail fins, and then move onto land. Other "neotenous" individuals retain their larval structures and continue an aquatic existence, overwintering in that form. Sexual maturity is reached the following spring, and mating begins again.

# ⑪ Profiles of Selected Prairie Reptiles

**Common Snapping Turtle**
*Chelydra serpentina*
*Identification:* Snapping turtles are unmistakable; the large head, long tail, and powerful front and hind legs set them apart from other Nebraska turtles. The triple-keeled rather than smoothly curved carapace is unique, and its posterior portion has uniquely serrated edges. Snapping turtles are the largest turtles in the state, with some individuals weighing up to about 50 pounds.

*Status:* Snapping turtles are widespread in Nebraska, and although they have not been reported from every county, they are likely to be found in almost any river or wetland in the state. Snapping turtles are most often found in shallow bodies of water, especially where aquatic vegetation or debris is present.

*Habitats and Ecology:* Snapping turtles move to land to deposit their eggs, the females digging out holes in areas of sandy soil and open vegetation. The sandy substrate evidently provides the best incubation conditions, and the short vegetation makes it easier for the hatchlings to disperse from the nest site. In Nebraska nesting begins in May and mostly occurs over a three-week period. Clutches of up to as many as 109 eggs have been found,

but they average about 50 eggs. The female does not protect them, and raccoons, skunks, and minks are all major egg predators. Hatching occurs from as early as August until as late as October. The sex of the hatchlings depends on incubation temperature, with males being produced at intermediate temperatures and females at both high and low temperatures. From 10 to 12 years are required to reach sexual maturity. Snapping turtles must be treated with great care when closely approaching them or lifting them by their carapace; their long necks can extend surprisingly far forward and backward, and they can easily bite off a finger. I once found an incapacitated western grebe with one missing leg that had obviously been cleanly sheared off by a snapping turtle. These turtles are sometimes hunted for food and are often killed simply because of their threat potential.

**Northern Painted Turtle**
*Chrysemys picta*
*Identification:* The painted turtle is well named: its plastron is reddish with an irregularly shaped black to brown central figure, and some red color is often present along marginal scutes of the carapace. (In the similar Blanding's and false

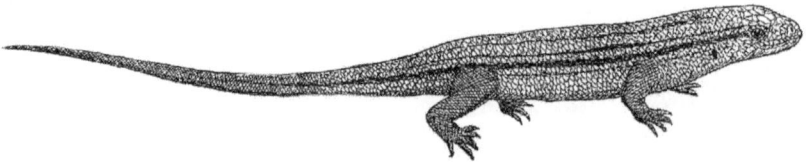

*Fig. 26. Northern prairie skink*

map turtles, the plastron is yellow with varying amounts of black markings.)

*Status:* This species is the commonest and one of the most widespread turtles in Nebraska, with records for all but 32 counties. It tends to favor large ponds and lakes over small ponds and shallow marshes.

*Habitats and Ecology:* Painted turtles can be found in larger, permanent ponds, lakes, and streams and larger slow-flowing rivers. Basking sites are an important part of the species' habitat, and individuals sometimes fight over preferred locations. The turtles are basically diurnal, with most foraging occurring in late morning and late afternoon, but nighttime activities have also been reported. They perform limited movements between ponds and have been found able to return home after displacements of up to 100 meters, evidently using a sun-compass guide. During winter they might remain active under the ice or burrow into mud at the bottom of a wetland. Mating occurs during early spring and fall, and females lay clutches of about 13 to 14 eggs, with as many as three clutches produced per year in the Nebraska Sandhills. Warmer nests produce young that are all or mostly females, colder nests all males, and intermediate nests produce a mixture of the sexes. Apparently the temperature present during the middle part of the incubation period determines sex. In western Nebraska the hatchlings spend their first winter in the nest and can survive freezing for at least 48 hours. Adult painted turtles eat larger aquatic invertebrates such as crayfish, insects, and insect larvae. They also consume small vertebrates, including salamanders, frogs, and fish, and plant materials as well. Iverson and Smith (1993) have described this species' natural history in Nebraska.

## Ornate Box Turtle
*Terrapene ornata*

*Identification:* Box turtles are easily identified—they are Nebraska's only entirely terrestrial turtle. They have a hinge on the plastron that allows its front end to be lifted and the turtles' head withdrawn protectively behind the high-domed carapace. Although the Blanding's turtle has a similar hinge and movable plastron, it is apparently not so highly effective in protecting the head. The box turtle has a mixture of yellow lines and spots radiating outward from the center of the carapace, and the plastron is dark brown

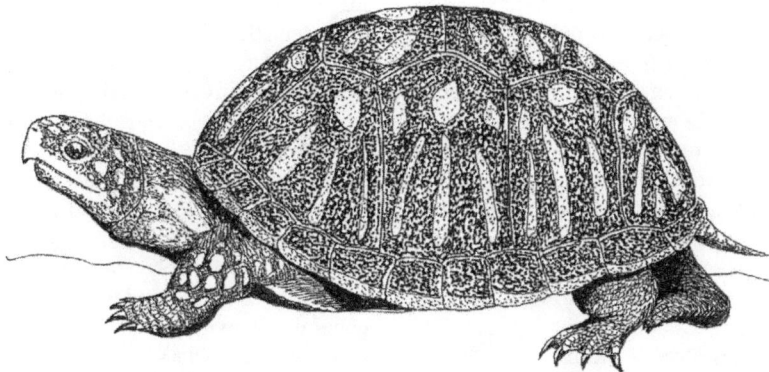

*Fig. 27. Ornate box turtle*

with yellow spots, whereas the Blanding's similarly dark carapace is sprinkled with small yellow spots, and the plastron is yellow with black smudges. The iris color is red in adult males, and varies from green to yellow-brown or maroon in females.

*Status:* The box turtle has been reported from nearly all the Panhandle counties but from relatively few counties in the eastern third of Nebraska. There is at least one Lancaster County record, but Spring Creek Prairie probably does not provide suitable sandy habitat. Still common in the Nebraska Sandhills, box turtles tend to wander during spring and summer—males probably in search of females—and have been proven to be able to find their way back to their home range after being displaced over distances of up to nine kilometers. During their travels the turtles often cross roads and frequently are accidently (or purposefully) run over by motorists. There are few natural predators, however, and Sandhills box turtles have been found to survive for up to at least 29 years. Legler (1960) has described this species' natural history.

*Habitats and Ecology:* Box turtles become active as early as mid-April in Nebraska, when males begin to seek out females. Mating extends from late April to early June. Clutch sizes are small in box turtles, averaging only about 4 or 5 eggs, but females probably produce more than one clutch per year in some populations. Like many other turtles, sex determination apparently depends on temperatures during the incubation period. Hatchlings overwinter in burrows below their nests in western Nebraska, and during hot weather adults tend to be inactive and remain in burrows.

## Northern Prairie Skink

*Plestiodon septentrionalis*
*Identification:* The northern prairie skink is the commonest (but rarely seen) lizard

in eastern Nebraska, its range extending west to Rock and Phelps Counties, and it is one of only two lizards reported in Lancaster County. It has not been reported from Spring Creek Prairie, although its Nebraska range corresponds closely with that of the historic tallgrass prairie. The other regional species, the five-lined skink, has five white lines extending down its back, whereas in the prairie skink the light mid-dorsal line is replaced by a broad brown stripe. In both species the chin and neck become bright reddish orange during the breeding season, and in both the tail is bright blue in juveniles. Fitch (1954) detailed the life history and ecology of the five-lined skink in Kansas.

*Status:* The secretive nature of this species makes its status difficult to establish, but it is often associated with open grasslands where flat rocks and loose soil allow it to dig an extensive tunnel system. It also occurs along sandbanks and on gravelly glacial outwashes.

*Habitats and Ecology:* This skink overwinters below the frost line under stones and debris, sometimes in large groups up to six feet below the surface. Skinks emerge from mid-April to early May, with mating occurring in late May. Females lay clutches of 4 to 18 eggs, averaging 11, and brood them until hatching occurs 2 to 4 weeks later. Females actively tend their nest to maintain proper environmental conditions for the eggs. They also remove infertile eggs and eggs with dead embryos that have become rotten. Like other skinks, juveniles have long, bright blue tails that can easily break off, should they be attacked by a predator. This survival strategy leaves the predator with only a tail and allows the victim to escape and eventually regenerate a new tail. Prairie skinks consume insects, spiders, and the larvae of various arthropods. Grasshoppers and other orthopterans are also an important part of their diet.

## Bullsnake

*Pituophis catenifer*

*Identification:* The bullsnake is Nebraska's most common large snake, with older adults often reaching 60 to 70 inches in length, and rarely approaching 100 inches. Its head is large and somewhat triangular, with alternating dark and yellow bands extending from the eye to the jaw, and a brown band extending diagonally from behind the eye to the angle of the jaw. More than 40 black to reddish brown blotches are scattered along the back, sides, and tail, and the yellow belly is also spotted with black. The dorsal scales are strongly keeled.

*Voice:* Like other snakes, bullsnakes are voiceless, but when threatened they can make a hissing sound by exhaling and passing air over a unique laryngeal structure. They also often shake their tail, which might resemble the rattle of a rattlesnake as it brushes through vegetation. A threatened bullsnake can even change the shape of its head to become more triangular, something like a rattlesnake's. Such features suggest that bullsnakes may be mimicking rattlesnakes as a protective adaptation (Kardong, 1980).

*Status:* Vying with the common gartersnake for being Nebraska's most widely distributed snake, bullsnakes have been recorded statewide except for 22 of Nebraska's east-central counties. They are also among the state's largest snakes.

*Habitats and Ecology:* This is primarily a grassland snake. It especially favors taller prairies but also often enters farmlands and urbanized areas; it rarely occupies woodlands. Bullsnakes are diurnal and are active from about April to October. They breed in spring, shortly after emerging from their winter burrows, and females lay eggs until as late as July. In Nebraska, clutches average 12 to 13 eggs but range from 8 to 17. The incubation period is about 70 days. At night the snakes return to their burrows, which are often gopher burrows (thus the common name variant of "gophersnake") or prairie dog burrows. They are also able to dig soil themselves by using head movements. Bullsnakes eat a variety of mammals (especially rodents), lizards, and ground-nesting and cavity-nesting birds as well as their eggs and nestlings. They are remarkably adept at climbing trees and can even climb vertical concrete walls to reach the nests of culvert-nesting cliff swallows.

## Common Gartersnake and Plains Gartersnake

*Thamnophis sirtalis* and *T. sirtis*

*Identification:* The common gartersnake closely resembles the equally common and even more widespread Plains gartersnake. Both species are strongly striped dorsally and along their sides with dark and light stripes. The common gartersnake alternates brick red and black vertical bars and wedges along its entire body, except for an unmarked white to bluish underside, whereas most Nebraska populations of the Plains gartersnake alternate yellow and black patterning. However a variant Plains gartersnake in eastern Nebraska also alternates red and black dorsally, much like the common gartersnake's markings, but its underside is spotted with black, rather than being unmarked. Both species have a conspicuous pale stripe down the middle of the back, which is yellow in the common gartersnake but usually bright orange in the Plains gartersnake.

*Status:* The common gartersnake has been recorded in all but 22 of the state's counties, and the Plains gartersnake in all but 10. Both species tend to favor locations close to water, but the common gartersnake seems to also have a preference for grassland-woodland transitional habitats where rocks and other escape cover are present.

*Habitats and Ecology:* Gartersnakes are diurnal and seasonally active in Nebraska, emerging in March from underground hibernacula that are shared by a large number of individuals, sometimes including other species of snakes. They return to their winter dens in late October but may briefly emerge during warm winter days. With spring emergence, mating begins (Joy and Crews, 1985). Males emerge from the hibernacula well before the females and then patrol the area while waiting for the females to emerge. From pheromone (cloacal secretion) trails left by females, the males can recognize those from the male's own den and may prefer to court them. Males may also use visual clues in their choices of females to be courted. Writhing knots of several simultaneously mating snakes can sometimes be seen. Female common gartersnakes lay clutches that vary from about 13 to 40 eggs; Plains gartersnake clutches average about 17 or 18 eggs in Nebraska. The young of both species hatch in July or August, and most adults and young both move into hibernacula by October. During summer, common gartersnakes are surprisingly mobile, with males having a home range of up to 35 acres, and females up to nearly 23 acres. Displaced common gartersnakes are able to orient themselves in the correct homeward direction, apparently using the sun as a compass. Adult common gartersnakes prey on a variety of animals, especially leopard frogs but also small rodents, earthworms, and many other prey, evidently using chemical clues to identify potential prey. Plains gartersnakes similarly mostly feed on amphibians, fish, earthworms, and slugs. Various studies indicate the annual survival rates of gartersnakes range from 34 to 50 percent, meaning that most individuals don't live beyond 3 to 4 years.

## Massasauga
*Sistrurus catenatus*

*Identification:* This rattlesnake and all three other species of Nebraska's venomous snakes can be identified as such by their vertically oriented pupils and by the presence of paired sensory pits located immediately behind their nostrils. They also all have wide, rather triangular-shaped heads, although this shape is not so obvious in the massasauga. The massasauga can be further recognized by the presence of nine unusually large scales on the top of its head; only three scales separate the eyes. The overall color of the snake is light brown to gray, with dark brown blotches on the back and sides, which become vertical bands toward the tail. A broad dark brown streak though the eye is bordered below by a broad white line, which loops down and encircles a second brown band. The tail rattles are relatively small in this species, which ranges from about 25 to 30 inches in length as adults, with females averaging longer and heavier than males. Unlike most Nebraska snakes, the massasauga is viviparous, giving birth to living young.

*Status:* Now extirpated from Lancaster County, this snake was reported from Nine-Mile Prairie in the 1940s by George Hudson. The massasauga has now been reduced to surviving apparently only in Pawnee County, although it is possible that relict populations still exist in remnant tallgrass prairies of southeastern Nebraska.

*Habitats and Ecology:* Massasaugas are active in Nebraska from April to October, in habitats varying from prairies to woodlands. In Iowa they have been reported to move to marshes to hibernate. Mating occurs during both spring and fall, with females in Missouri producing young on alternate years. Litter sizes in one Colorado study ranged

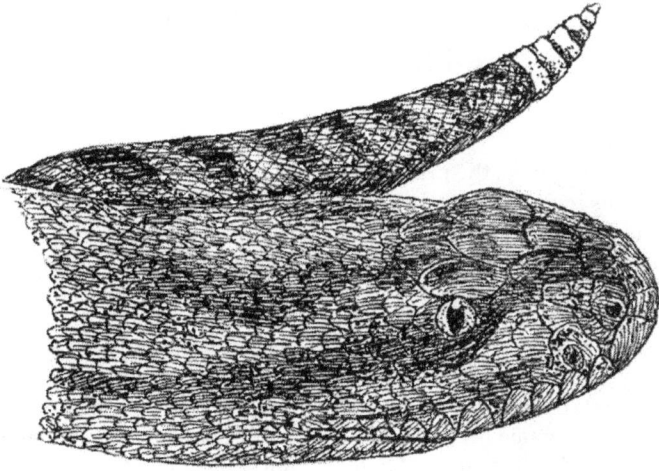

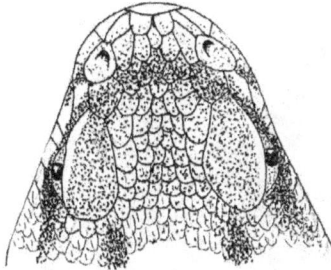

*Fig. 28. Massasauga*

from 5 to 7 young but averaged only 4.5 in another Colorado study. A wide diversity of prey is consumed, including lizards, snakes, frogs, and bird eggs as well as small mammals and invertebrates. Massasaugas are rare in Nebraska but are not considered threatened. Being bitten by one is serious but not fatal, as rather small amounts of venom are injected.

# 12 The Butterflies of Spring Creek Prairie and Lancaster County

Except for the skippers, the members of the families included in this chapter are typical butterflies, consisting mostly of fairly large insects with large, brightly colored, highly patterned wings; long sucking mouthparts; and antennae that terminate in an enlarged knob-like shape. They are active during the day (diurnal) and rely on their color vision to find flowers that contain nectar (plants that have evolved colors, shapes, and scents that attract pollinating insects). Many of the members of the four families in this list are species whose adults vary enough in the color, patterning, and shape of their wings to allow for ready field identification.

Ultraviolet (UV) perception is common in butterflies, which allows them to see otherwise invisible nectar guides on flowers. These guides provide them with a communication channel that is in-

*Fig. 29. Monarch butterfly: mature pupa, empty chrysalis, and adult*

visible to most predators but facilitates both sex recognition and mating behavior. On finding a female, a male butterfly might hover over her or beside her, and if the female remains still he will attempt to copulate. Copulation in butterflies often lasts for an hour or more.

In contrast to the diurnal butterflies, moths are usually nocturnal and rely more on scent than vision to find nectar-bearing flowers and locate mates. (Color vision is probably weak or lacking in the nocturnal species.) Furthermore, moths are rarely brightly colored and have long, often branched, and highly odor-sensitive antennae. The flowers that they pollinate are typically night blooming, usually white or pale in color, and highly odorous. The larvae of both butterflies and moths ("caterpillars") are voracious leaf eaters. Larvae construct often distinctive pupae (chrysalids or cocoons) prior to becoming dormant and metamorphosing into winged adults.

This Lancaster County butterfly checklist is based in part on Daggert et al. (2005); the species reported from Spring Creek Prairie are based on an in-house checklist and are shown in **bold.** Measurements refer to the distance between the tips of the horizontally spread wings. In this book, "migrant" refers to a southern species that shows up in Nebraska but does not overwinter here. "Widespread" indicates that Lancaster County lies well within the species' generally broad Nebraska range, and local sightings are not unusual. The sequence of species and larger groups here follows the popular butterfly field guide by Brock and Kaufman (2006); the page reference included with each species' listing refers to the species' descriptive account in that book. The R. T. Peterson–style field guide by Opler and Malikul (1998) is also excellent, with paintings and maps of all butterfly and skipper species that range west to central Nebraska. A new superb guide to North American butterflies by Glassberg (2017) has more than 3,500 color photos. An online checklist (with species lists for individual states) with photos and detailed descriptions of North American butterfly and moth species is available at https://www.butterfliesand-moths.org/checklists, the website of the Butteflies and Moths of North America (BAMONA) project.

## Swallowtails and Parnassians (Papilionidae)

The Papilionidae family of 600 to 700 species of generally large butterflies occurs worldwide but is mostly tropical. Many of the species, but not all, are "swallow-tailed"—namely, they have one or two extended "tails" on their hindwings. One group, the parnassians (which was named after the Greek mountain Parnassus, which was sacred to Apollo and the Muses), lack "tails" and brilliant colors and occur mainly in the western mountains; all of the Nebraska species are true swallowtails. When feeding, swallowtails often rapidly beat their wings, probably because their large size might cause slender flowerheads to bend under their weight. Swallowtail eggs are spherical and pale green, and the young caterpillars often resemble bird droppings. Older papilionid caterpillars are of different colors and sometimes have large eyespots that serve as a visual defense by alarming possible predators. Swallowtail caterpillars can also emit a sickening odor as a chemical defense. Their pupae are well camouflaged, attached to branches with silk supports.

**Eastern tiger swallowtail,** *Papilio glaucus.* 2.5–4.5 inches. Widespread; p. 20.

Zebra swallowtail, *Eurytides marcellus.* 1.9–3.0 inches. Edge of range; p. 24.

**Pipevine swallowtail,** *Battus philenor.* 2.7–4 inches. Migrant; p. 26.

**Black swallowtail,** *Papilio polyxenes.* 2.7–4 inches. Widespread; p. 28.

Giant swallowtail, *Papilio cresphontes.* 3.7–5 inches. Widespread; p. 38.

## Whites, Sulfurs, and Yellows (Pieridae)

This very large family of butterflies, the Pieridae, includes 1,100 to 2,000 species, most of which are tropical, but some reach the Arctic Circle and alpine environments in their distribution. Many are yellow ("sulfurs") and historically responsible for the common name "butterfly." Others have white or orange wings sparsely marked with black. The "whites" have white wings with black markings, the "marbles" and "orange tips" often have orange wingtips and green marbling on their undersides, and the "sulfurs" are mostly yellow but sometimes one sex is white. Besides sexual appearance differences, there are also seasonal differences in some species. Many species appear differently under ultraviolet (UV) light, which butterflies can perceive, and such features are evidently important in species and sexual recognition. Males also emit scents (pheromones) to attract females. All adult pierids feed on flower nectar, and males often gather at mud puddles. The eggs of this family are spindle shaped, and their pupae are suspended by silk supports. The larvae often eat plants of the legume or mustard family; the sulfur species store mustard oil in their tissues and may be distasteful to predators.

### Whites and Marbles
*(1.2–1.8 inches, wings white, variably black-spotted)*

**Cabbage white,** *Pieris rapae.* 1.25–1.8 inches. Widespread (introduced); p. 46.

**Checkered white,** *Pieris protodice.* 1.25–1.8 inches. Widespread; p. 48.

Olympia marble, *Euchloe olympia.* 1.2–1.7 inches. Widespread, p. 56.

### Sulfurs and Yellows
*(1–2.7 inches, wings yellow, orange, or white)*

**Orange (Alfalfa) sulfur,** *Colias eurytheme.* 1.3–2.3 inches. Widespread, p. 60.

**Clouded (Common) sulfur,** *Colias philodice.* 1.1–2.0 inches. Widespread, p. 60.

Southern dogface, *Zerene caesonia.* 1.7–2.5 inches. Migrant; p. 68.

**Little yellow,** *Pyrisitia lisa.* 1.1–1.7 inches. Widespread; p. 70.

**Dwarf yellow (Dainty sulfur),** *Nathalis iole.* 0.7–1.3 inches. Widespread; p. 70.

**Cloudless sulfur,** *Phoebis sennae.* 1.8–2.7 inches. Near edge of range; p. 74.

Sleepy orange, *Abaeis nicippe.* 1.4–2.3 inches. Migrant; p. 68.

## Gossamer-winged Butterflies (Lycaenidae)

This huge family of generally small butterflies, the Lycaenidae, consists of 4,000 to 6,000 species worldwide, of which about 100 occur in North America. Four groups (three of which are represented locally) occur in North America; one group (the blues) includes the continent's smallest butterfly. All members of the family are characterized by the fact that, in males, the front pair of legs are reduced in size and are nonfunctional for walking. Nearly all the species are highly colorful, with both pigmented scales (brown and gray) and iridescent scales (green, blue, purple, and bright reddish orange). Their eggs are flattened and sculpted in various ways, some resembling turbans. The larvae of lycanids are usually green or brown and covered with fine hairs. In some species the larvae produce a sugary "honeydew" and are tended like cattle by ants. In many species the pupa will suddenly flex if disturbed, producing an unexpected movement and faint sound that might help deflect predators. Most of the species perch with their wings tightly closed, but they can often still be identified by their underwing patterns.

### Coppers and Harvester
*(0.8–1.7 inches; wings with copper and black patterns)*

**Harvester,** *Feniseca tarquinius*. 0.85–1.3 inches. Edge of range; p. 80.

**Gray copper,** *Lycaena dione*. 1.5–1.7 inches. Widespread; p. 82.

**Bronze copper,** *Lycaena hyllus*. 1.2–1.6 inches. Widespread; p. 88.

### Hairstreaks
*(0.8–1.3 inches; hindwings with threadlike tails)*

**Gray hairstreak,** *Strymon melinus*. 1.1–1.4 inches. Widespread, p. 92.

Banded hairstreak, *Satyrium calanus*. 0.8–1.3 inches. Near edge of range; p. 94.

Coral hairstreak, *Satyrium titus*. 1–1.5 inches. Widespread; p. 98.

Acadian hairstreak, *Satyrium acadia*. 1.2–1.4 inches. Widespread; p. 98.

Striped hairstreak, *Satyrium liparops*. 1–1.5 inches. Widespread; p. 94.

**Henry's elfin,** *Callophrys henrici*. 1.1–1.25 inches. Edge of range; p. 110.

### Blues
*(0.6–1.2 inches; tiny, wings mostly sky blue above)*

**Eastern tailed-blue,** *Cupido comyntas*. 0.6–1 inch. Widespread; p. 124.

Marine blue, *Leptotes cassius*. 1–1.1 inches. Edge of range; p. 126.

**Reakirt's blue,** *Hemiargus isola*. 0.6–1 inch. Widespread; p. 128.

**Summer azure,** *Celastrina neglecta*. 0.7–1.2 inches. Widespread; p. 130.

**Melissa blue,** *Plebejus melissa*. 1–1.4 inches. Widespread; p. 134.

Lupine blue, *Plebejus lupini*. 0.9–1.1 inches. Edge of range; p. 132.

### Brush-footed Butterflies (Nymphalidae)

The large Nymphalidae family of butterflies is highly diverse and in some ways represents a leftover group of species that don't seem to fit anywhere else. Some of the included species—such as the snout butterflies, woods and satyrs, and milkweed butterflies—are sometimes classified as separate families. However, all the species included here have a front pair of legs that is reduced in size and covered by long hair-like scales. Typical nymphalids vary greatly in size and shape but are mostly medium to large, colorful butterflies that fly fast. They are often varied in their wing shapes, such as in having irregular wing margins or tail-like extensions on the hindwings. Although most are attracted to flowers, some species—such as the admirals and the hackberry butterfly—feed on tree sap, rotting fruit, dung, or carrion. Their eggs are ribbed and laid singly or in groups. Their caterpillars typically are covered with spines or bristles, and some have paired "horns" on their heads. Their pupae hang downward from silken mats. Among the aberrant groups are the snout butterflies (one species in Nebraska), in which the labial palps are extended forward, resembling a long proboscis, perhaps to mimic a leaf petiole and enhance the otherwise leaf-like appearance of the perched butterfly. Another divergent group includes the satyrs, wood nymphs, and browns. Most of these butterflies are a patterned brown with marginal eyespots on the wings, especially on the undersides. Males of many species in this group have scent glands on their wings, and swollen vein bases on the forewings are hearing structures for predator detection. Lastly, the milkweed butterflies are a group of mostly tropical species that includes the monarch and queen. They are unique in that they have unscaled antennae, and males have pheromone-producing structures on their hindwings and abdomens. Their larvae feed on milkweeds or dogbanes, both of which are sources of poisonous cardiac glycosides that are stored in their bodies and make them distasteful if not poisonous to predators. The books *Butterflies of Houston and Southeast Texas* (Tveten and Tveten, 1996) and *Butterflies of North America* (Scott, 1992) provide a wealth of information on both adult and larva butterflies and their often amazing life histories.

### Fritillaries
*(1.4–3.8 inches; wings mostly rusty orange and black)*

**Variegated fritillary,** *Euptoieta claudia.* 2.5–2.6 inches. Widespread, p. 156.

**Great spangled fritillary,** *Speyeria cybele.* 2.6–3.5 inches. Widespread; p. 158.

**Regal fritillary,** *Speyeria idalia.* 2.9–3.8 inches. Species of concern (US); p. 158.

**Silver-bordered fritillary,** *Boloria selene.* 1.4–2.0 inches. Widespread; p. 170.

### Crescents and Checkerspots
*(1–1.9 inches; wings spotted orange and black)*

**Pearl crescent,** *Phyciodes tharos.* 1–1.6 inches. Widespread; p. 177.

**Silvery checkerspot,** *Chlosyne nycteis.* 1.3–1.9 inches. Widespread; p. 184.

**Gorgone checkerspot,** *Chlosyne gorgone.* 1–1.7 inches. Widespread; p. 184.

### Snouts
*(1.2–1.9 inches; head with snout-like extension)*

**American snout,** *Libytheana carinenta.* 1.2–1.9 inches.Widespread; p. 222.

### Typical Brushfoots
*(1.8–3.3 inches; wings with diverse colors and patterns)*

**Mourning cloak,** *Nymphalis antiopa.* 2.5–3.3 inches. Widespread; p. 202.

Buckeye, *Junonia coenia*. 1.5–2 inches. Widespread; p. 206.

Viceroy, *Limenitis archippus*. 2.3–3.1 inches. Widespread; p. 210.

Goatweed leafwing, *Anaea andria*. 1.8–2.5 inches. Widespread; p. 220.

### *Commas and Question Marks*
*(1.6–3.1 inches; wing edges sharply angled)*

Question mark, *Polygonia interrogationis*. 1.2–2.5 inches. Widespread; p. 196.

Eastern comma, *Polygonia comma*. 1.6–2.3 inches. Widespread; p. 196.

Gray comma, *Polygonia pogne*. 1.6–3.1 inches. Widespread; p. 198.

### *Ladies and Admirals*
*(1.6–3.5 inches; wings often white-banded)*

Red admiral, *Vanessa atalanta*. 1.6–2.3 inches. Widespread; p. 202.

Painted lady, *Vanessa cardui*. 1.7–2.6 inches. Widespread; p. 204.

American lady, *Vanessa virginiensis*. 1.6–3.1 inches. Widespread; p. 204.

Red-spotted purple, *Limenitis arthemis astyanax*. 2.3–3.5 inches. Widespread; p. 210.

### *Emperors*
*(1.5–2.4 inches; forewings with many white eyespots)*

Hackberry emperor, *Asterocampa celtis*. 1.5–2.2 inches. Widespread; p. 222.

Tawny emperor, *Asterocampa clyton*. 1.5–2.4 inches. Eastern Nebraska; p. 222.

### *Milkweed Butterflies*
*(2.7–3.9 inches; wings black-banded orange)*

Monarch, *Danaus plexippus*. 2.7–3.9 inches. Widespread; p. 226.

### *Satyrs*
*(1.3–2.7 inches; brown, forewings with dark eye-spots*

Little wood-satyr, *Megisto cymela*. 1.3–1.7 inches. Widespread, p. 230.

Common wood-nymph, *Cercyonis pegala*. 1.7–2.7 inches. Widespread; p. 236.

Northern pearly-eye, *Enodia anthedon*. 2.1–2.6 inches. Widespread; p. 238.

## Skippers (Hesperiidae)

The Hesperiidae family is a broadly distributed group of about 1,000 species worldwide. Skippers are generally small, inconspicuous butterflies, most of which are very hard to identify in the field and thus tend to be overlooked. They have fast, darting flights (thus the name "skipper") and are unique in that adults of most of the species have antennae with slender, recurved tips rather than the club-like tips of typical butterflies. Many hesperid species feed on grasses as larvae (the grass skippers), and the adults of this group rest with their forewings raised and their hindwings somewhat spread, producing a distinctive "jet-plane" profile. Many species in the grass skipper group have yellow to orange wings with a dark area of scent scales on the upper side of each forewing, which is called the stigma.

### Spread-wing Skippers
*(1.0–2.5 inches; wings held horizontally at rest)*

**Silver-spotted skipper,** *Epargyreus clarus.* 1.6–2.5 inches. Widespread; p. 256.

Funereal duskywing, *Erynnis funeralis.* 1.3–1.7 inches. Migrant; p. 280.

**Wild indigo duskywing,** *Erynnis baptisiae.* 1.4–1.6 inches. Widespread; p. 286.

**Common checkered-skipper,** *Pyrgus communis.* 1–1.3 inches. Widespread; p. 288.

**Hayhurst's scallopwing,** *Staphylus hayhurstii.* 1–1.2 inches. Widespread; p. 296.

**Common sootywing,** *Pholisora catullus.* 1–1.2 inches. Widespread; p. 298.

Roadside skipper, *Amblyscirtes vialis.* 0.9–1.3 inches. Widespread; p. 342.

### Grass Skippers
*(0.9–1.8 inches; forewings variably raised at rest)*

**Sachem,** *Atalopedes campestris.* 1.4–1.6 inches. Widespread; p. 302.

Fiery skipper, *Hylephila phyleus.* 1.25–1.5 inches. Edge of range; p. 302.

**Least skipper,** *Ancyloxypha numitor.* 0.9–1.2 inches. Widespread; p. 304.

Ottoe skipper, *Hesperia ottoe.* 1.4–1.7 inches. Local, species of concern (Nebraska); p. 318.

**Peck's (Yellowpatch) skipper,** *Polites peckius.* 1–1.25 inches. Widespread; p. 322.

**Tawny-edged skipper,** *Polites themistocles.* 0.9–1.5 inches. Widespread; p. 324.

Northern broken-dash, *Wallengrenia egeremet.* 1–1.5 inches. Edge of range; p. 326.

Dun skipper, *Euphyes vestris.* 1.2–1.35 inches. Widespread; p. 326.

**Little glassywing,** *Pompeius verna.* 1–1.5 inches. Edge of range; p. 326.

**Hobomok skipper,** *Poanes hobomok.* 1.4–1.7 inches. Widespread; p. 330.

Zabulon skipper, *Poanes zabulon.* 1.4–1.6 inches. Widespread; p. 330.

Iowa skipper, *Atrytone arogos iowa.* 1.1–1.8 inches. Local; species of concern (Nebraska); p. 336.

**Delaware skipper,** *Anatrytone logon.* 1–1.7 inches. Widespread; p. 336.

Eufala skipper, *Lerodea eufala.* 1.2–1.4 inches. Edge of range; p. 352.

# ⚙️ The Moths of Lancaster County and Eastern Nebraska

This list contains only some of the larger and more conspicuous moths likely to be seen at Spring Creek Prairie. No list of moths is available specifically for Spring Creek; the following refers to all of Lancaster County. A longer list of Nebraska's moths can be found in *The Nature of Nebraska* (Johnsgard, 2001b). Nebraska's beautiful underwing moths were illustrated by Jordison (1996). Many of the moth species listed here are described and illustrated in *A Field Guide to Moths of Eastern North America* (Covell, 1984). Plate numbers in that book are indicated in this list as "Pl." Any page references ("p.") refer to *Butterflies and Moths of Missouri* (Heitzman and Heitzman, 1996). Species illustrated in the *Peterson First Guide to Butterflies and Moths* (Opler, Wright, and Peterson, 1994) are marked with an asterisk (\*); those illustrated in *Insects in Kansas* (Salsbury and White, 2000) are marked with an @ symbol. Some of the sphinx moths are mobile and somewhat migratory; thus, out-of-range vagrants are not rare. Taxa are arranged alphabetically, initially by genus and secondarily by species. Measurements refer to distance between the tips of the horizontally spread wings.

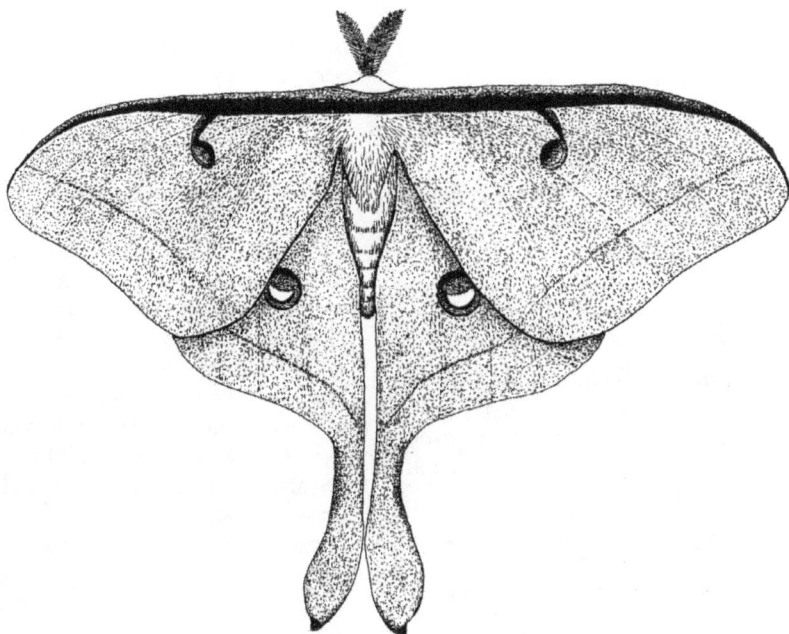

*Fig. 30. Luna moth, male*

## Saturniidae: Silk Moths

Silk moths are the largest and most beautiful of the Nebraska moths, a group noted for their large size, wing patterns that often include large eye-spots, and larvae that spin silken cocoons to house and protect the pupating larvae as they undergo metamorphosis into flying adults. Adult males have extremely large antennae that are highly sensitive to species-specific scents (pheromones) emitted by adult females; the scents can be detected from miles away by males. The mouthparts of adults are vestigial—adults often live for less than two weeks after emerging from their cocoons. All the Nebraska species overwinter as pupae. Tuskes, Collins, and Tuttle (1996) have monographed all 127 of the North American species of silk moths, and Cody (1996) has provided magnificent paintings of many. Ratcliffe (1993) has described the Nebraska species.

### BUCK AND IO MOTHS, SUBFAMILY HEMILEUCINAE

Io moth, *Automeris io*. 2–3.25 inches. Widespread; PI 1, 2, 10.*@

Polyphemus moth, *Antheraea polyphemus*. 3.5–5.5 inches. Widespread; PI 1, 2, 9.*@

### GIANT SILKWORM MOTHS, SUBFAMILY SATURNIINAE

Luna moth, *Actias luna*. 3–4.25 inches. Eastern counties. PI 1, 2, 9.@
Cecropia silkmoth, *Hyalophora cecropia*. 3.75–6 inches. Widespread; PI 1, 2, 10.*@

### ROYAL MOTHS, SUBFAMILY CITHERONINAE

Rosy maple moth, *Dryocampa rubicunda*. 1.25–2.25 inches. Eastern counties. PI 8.*

Bicolored honey locust moth, *Sphingicampa bicolor*. 2–2.85 inches. Eastern counties; PI 8.@

Bisected honey locust moth, *Sphingicampa bisecta*. 2–2.85 inches. Eastern counties; PI 8.*

## Sphingidae: Sphinx Moths

Sphinx moths are so called because their disturbed larvae often assume a defensive stance with their head and thorax raised into a sphinxlike posture. They are also called "hawkmoths" or "hummingbird moths" and often are confused with hummingbirds. These are the jet planes of the moths, with long, tubular bodies and wingbeats so fast their wings become blurred to human eyes. They have long narrow forewings and much smaller and variably colorful hindwings. They are strong fliers and have been collected far out at sea, up to 600 miles from land. Although they have been clocked at speeds of up to 30 miles per hour, they are also able to hover motionless in front of a flower while foraging. However, while at rest the exposed wings, like the body, present a complex camouflaged pattern that usually closely resembles wood bark.

Besides having fine vision, sphinx moths also have excellent olfactory abilities, which help direct them to night-blooming plants rich in nectar. They are mostly nocturnal and crepuscular, with eyes that are mainly adapted for night vision, which probably extends into the ultraviolet (UV) range. Some sphinx moths are also diurnal, including two of Nebraska's 32 species that have transparent areas on their wings ("clearwings"); they somewhat resemble bumblebees when in flight. The nocturnal species are often attracted to electric lights at night. The proboscis of sphinx moths is extremely long and hollow, and in species where its length is more than about 100 mm the moth hovers hummingbird-like in front of the flower, while inserting its proboscis into it. One very large white Madagascan orchid (*Angraecum sesquipedale*) has a nectary so deep (10–11 inches) that, on seeing it, Charles Darwin famously speculated in an 1862 correspondence to an orchid fancier friend that there must be some unknown moth with a proboscis long enough to reach the nectar. It was not until 1903, more than two decades after his death, that the predicted insect, a large sphinx moth (*Xanthopan morganii praedicta*), was finally discovered, and not until 1992 that proof of Darwin's amazing prescience was proven photographically.

Tuttle (2007) has recently monographed all 127 North American hawk moth species. Messenger (1997) described all the Nebraska species in detail and photographically illustrated them in color.

Titan sphinx, *Aellopos titan*. 4.1–5.2 inches. Scattered records, tropical vagrant; Pl 5.*@

Pink-spotted hawkmoth, *Agrius cingulatus*. 4.1–5.2 inches. Migrant, gardens, scrub; Pl 3; p. 203.

Nessus sphinx, *Amphion floridensis*. 1.6–2.4 inches. Scattered records, woodlands, diurnal; Pl 6.

Elm sphinx, *Ceratomia amyntor*. 3.8–5 inches. Widespread, woodland edges; Pl 3.*@

Catalpa sphinx, *Ceratomia catalpae*. 2.8–4 inches. Scattered records, woodland edges; Pl 5.

Hagen's sphinx, *Ceratomia hageni*. 3.3–4 inches. Few eastern counties, woodland edges; Pl 4.*@

Waved sphinx, *Ceratomia undulosa*. 3.1–4.9 inches. Widespread, woodland edges; Pl 4; p. 206.

Virginia creeper sphinx, *Darapsa myron*. 1.7–2.8 inches. Scattered records, woodland edges; Pl 6.*@

Lettered sphinx, *Deidamia inscriptum*. 1.6–3.0 inches. Scattered records, woodland edges; Pl 7.*@

Achemon sphinx, *Eumorpha achemon*. 3.6–4.4 inches. Widespread, woodland edges and scrub; Pl 3.*@

Ello sphinx, *Erinnyis ello*. 3–3.3 inches. Two counties only, tropical vagrant; Pl 3.

Obscure sphinx, *Erinnyis obscura*. 2.4–2.8 inches. Few counties, southern vagrant; Pl 6.

Pandorus sphinx, *Eumorpha pandora*. 3.8–5 inches. Scattered records, woodland edges; Pl 3; p. 215.*@

Snowberry clearwing, *Hemaris diffinis*. 1.5–1.7 inches. Widespread, woods and gardens; Pl 6; p. 214.*@

Hummingbird clearwing, *Hemaris thysbe*. 1.7–1.8 inches. Few records, woods and fields; Pl 6; p. 213.

White-lined sphinx, *Hyles lineata*. 2.7–3.9 inches. Widespread, gardens and fields; Pl 3; p. 220.*@

Walnut sphinx, *Amorpha juglandis*. 1.9–3.3 inches. Scattered records, woodland edges; Pl 6; p. 211.

Five-spotted hawkmoth, *Manduca quinquemaculata*. 3.9–5.9 inches. Widespread, woodlands; Pl 3; p. 203.*@

Carolina sphinx, *Manduca sexta*. Scattered records. 4.4–5.4 inches. Scrub and gardens; Pl 1; p. 202.*@

Blind-eyed sphinx, *Paonias excaecatus*. 2.3–4.1 inches. Few counties, woodlands; Pl 6; p. 209.

Modest sphinx, *Pachysphinx modesta*. 3.3–5.5 inches. Wetlands; Pl 3; p. 212.*@

Small-eyed sphinx, *Paonias myops*. 1.9–3.2 inches. Scattered records, woodland edges; Pl 6; p. 210.*@

Plebeian sphinx, *Paratrea plebeja*. 2.4–3.3 inches. Few eastern counties, scrub and gardens; Pl 5.

Abbott's sphinx, *Sphecodina abbotti*. 2.5–3 inches. Southeastern counties; Pl 7.*@

Twin-spotted sphinx, *Smerinthus jamaicensis*. 2.2–3 inches. Widespread, damp areas; Pl 6; p. 208.*@

Great ash sphinx, *Sphinx chersis*. 3.4–5.6 inches. Widespread, woodlands; Pl 4.

Wild cherry sphinx, *Sphinx drupiferarum*. 3.6–4.8 inches. Few counties, woodlands; Pl 4; p. 207.

Laurel sphinx, *Sphinx kalmiae*. 3.2–4.5 inches. Eastern vagrant, woodland edges; Pl 4.

Vashti sphinx, *Sphinx vashti*. 2.9–4.3 inches. Widespread, scrub and gardens; Pl 4.

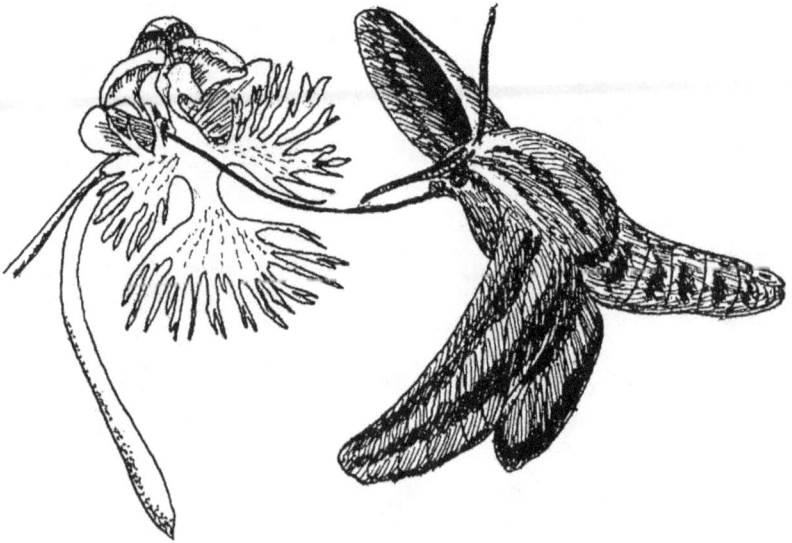

*Fig. 31. White-lined sphinx*

# 14 Profiles of Selected Prairie Butterflies and Moths

**Eastern Tiger Swallowtail**
*Papilio glaucus*
This beautiful butterfly is the most common of eastern Nebraska's swallowtails and is easily recognized by the four black "tiger" stripes on its yellow forewings. The hindwings each have a long black "tail," and females are very similar to males but have more extensive blue on the hindwings. Farther south in the Great Plains, an overall blackish color variant commonly occurs among females, but these are relatively rare in Nebraska. Although males prefer to mate with yellow females, the black ones are less vulnerable to attack by birds because they closely resemble the inedible pipevine and spicebush butterflies, which are both very common in the South but barely reach Nebraska in their range. Tiger swallowtails emerge from their winter chrysalids in April and are active until late September. The adults are commonly seen in gardens, where honeysuckles and milkweeds are favorite nectar sources. They also visit dung and carrion, and suck water from puddles. Common food plants of the larvae are apples, chokecherries, and other deciduous trees and shrubs. Their eggs are laid singly, often on ash trees, and the young larvae resemble bird droppings. Older larvae are leaf green, with two large, conspicuous, yellow and blue eyespots, plus a narrow white collar. During their long season of activity the adults produce two broods.

**Orange Sulfur**
*Colias eurytheme*
The orange sulfur butterfly, also often called the alfalfa butterfly, is one of Nebraska's most common butterflies, if not the most common. The males are orange above with wings bordered broadly in black, and there is a small black spot on each forewing. Females occur in two forms. One is like the male, except the black forewing borders have small orange spots. The other form has greenish white wings but otherwise has the same general pattern as the yellow form. The wings of both sexes are greenish yellow below with a pink-edged silver spot on each hindwing. Males prefer to mate with the orange females, but the white ones produce offspring more rapidly, which is an advantage in the northern Plains where the season is short. In the southern states the yellow form is more successful reproductively and is more common. Additionally, ultraviolet (UV) reflectance from the wings is important in mate selection. Only males reflect UV light from their upper wing surfaces, and females will mate only with males that reflect that light pattern. However, males are attracted to the yellowish green color on the undersides of female wings and are repulsed by UV reflections (Scott, 1992). Sulfur butterflies are active from May to November, overwintering as a chrysalis. Probably three broods are normally produced each season. Adults feed on a wide variety of plants and often gather at mud puddles. Their host plants also vary, but alfalfa and white clover are favored. The caterpillar is dark green with white side stripes outlined in black.

**Regal Fritillary**
*Speyeria idalia*
The regal fritillary is one of Nebraska's most beautiful butterflies, and one of our

most rapidly declining species, as well as one closely associated with native prairies. It has disappeared from its range east of the Mississippi River and now mainly occurs on native grasslands between Kansas and North Dakota. Like other large fritillaries it is mostly reddish orange with black spots and bars. Its hindwings are velvety black, with two rows of lighter spots. Both rows are creamy white in females, but in males the outer row is orange. Females are also slightly larger than males and are a richer shade of reddish orange. Both sexes have distinctive, large, black-edged, and somewhat triangular white spots on the rear underwings. This unique spotting pattern distinguishes them from the rather similar great spangled fritillary, and from all other fritillaries except the rare Gulf fritillary (*Agraulis vanilla*). Adults are active from early June to middle or late September, and they feed on a variety of prairie plants, including butterfly milkweed, coneflowers, gayfeathers, vervains, and thistles. Their eggs are deposited near violets of various species. The caterpillars are yellowish brown to black with dark markings and yellow stripes, as well as black-tipped branching white to yellowish dorsal spines. There is a single brood per season. After it hatches, the unfed larvae overwinter on the ground under leaves, which helps explain the late appearance of adults the following spring.

## Painted Lady
*Vanessa cardui*
Another of the very common butterflies of Nebraska, the painted lady is one of a group of three "lady" butterflies. They are all medium-sized orange and black butterflies with white-spotted black wingtips; they have from two to five eyespots on the undersides of the hindwings, which are otherwise mostly patterned with brown and white. The

painted lady has four or five such eyespots, compared with two in the American lady, a less common species in Nebraska. Most painted ladies that breed in Nebraska probably overwinter in the southern states or northern Mexico. These migrate north in huge flocks, and in some years are so numerous as to reduce visibility and interfere with traffic. In such years the resulting breeding populations are enormous, such as in 2017, when it was often possible to see as many as a dozen butterflies on a single flowering shrub or forb. They usually arrive in Nebraska during April and usually have departed by mid-November. They are found in all open areas but favor pastures, fields, meadows, and gardens. They feed on a wide variety of flowers, such as composites, milkweeds, coneflowers, lilacs, asters, and zinnias, as well as on tree sap and rotting fruit. Their eggs are likewise deposited on a wide array of more than 100 species of host plants, commonly including legumes, mallows, and thistles. (The species' Latin specific name *cardui* refers to *carduus*, a thistle.) Their pale green eggs are laid singly on the tops of leaves, and the larva builds an individual nest by folding and binding a leaf together with silk. The larvae are spiny and vary greatly in color, except for their black heads. Apparently no painted ladies overwinter as chrysalids, but some adults overwinter as far north as Texas.

## Monarch
*Danaus plexippus*
Monarchs are almost everybody's favorite butterfly. They are large, common, and beautiful, and they annually perform amazing continental migrations, using navigation skills and endurance capabilities that stagger the imagination. They are large orange and black butterflies, with black veins and white-spotted wing borders. The same pattern

shows on their undersides but is more yellow than orange. The female is less bright than the male, has wider black veins, and lacks the black oval patch on each rear wing that marks the male's sex pheromone source. The viceroy butterfly, a species that gains some protection from potential predators by mimicking and being visually confused with the inedible monarch, is similar in color and pattern but has a black band that extends across the hind wing that is absent in the monarch. Larval monarchs consume the leaves of milkweeds, extracting and storing cardiac glycosides, rendering their bodies distasteful and emetic to birds, a protection that carries over through pupation and into adulthood. After wintering in the mountains of Michoacan, in central Mexico, millions of monarchs begin migration north in spring, arriving in southern Texas during March and reaching Nebraska by early May. This northward trip is achieved through a series of rapid reproductive cycles, so that by the time monarchs have reached the northern Great Plains states or southern Canada they are as many as seven generations removed from those that started out from Mexico. Their cone-shaped eggs are laid on the underside of milkweed leaves. When the larvae are given a choice, they select leaves with an intermediate level of glycosides, which will not kill them but will provide adequate protection from predators. The larvae are conspicuously ringed with yellow, white, and black bands A pair of fleshy tentacles is located on the thorax, and a second, shorter pair is situated near the end of the abdomen. The pupa is a beautifully shaped green ovoid cylinder with a ring of black and specks of metallic gold. When the last brood of the summer emerges in late August or September, individuals begin flying south. In Nebraska the last of the migrants might be seen into late October. The fall mi-

grants apparently rely instinctively on the sun for navigation, flying more than a thousand miles over land that was last seen by butterflies at least seven generations removed from the first spring migrants. Assuming that a monarch is about the same weight as a honeybee (about 0.11 gram or 0.00025 pound), and extrapolating that to the weight of a 150-pound human, the human weighs 600,000 times as much. If we extrapolate the roughly 1,500-mile flight a monarch makes to central Mexico from Nebraska by 600,000-fold to adjust for the weight difference, it would be the equivalent to a human-sized monarch flying 900 million miles, or nearly ten times greater than the distance from the earth to the sun!

## Luna Moth
*Actias luna*

The luna (its name is derived from Luna, the Roman goddess of the moon) is an ethereal moth that once seen is never forgotten. It is a bright, almost luminous green, with two elegant, long "swallow" tails that are often twisted at the tops, and a pair of brown-edged eye-like oval spots on the hind wings. A pair of similarly colored but smaller and more oval eye-spots are present on the forewings. These eye-spots are connected by dark stripes to a purplish-brown leading-edge wing stripe that extends along the length of each forewing. The moth's fur-like scaled body and head are nearly white, with two antennae that are large and feather-like in form, especially in males, which are slightly lighter in color than females. Males are highly sensitive to a sexual pheromone emitted by receptive females and are reputed to be able to home in on such females from several miles away. The pheromones are released at night, over a very brief period called the "calling time," which is at about midnight in the luna. The mouthparts of adults are nonfunctional, so no feeding can be done. Thus, the life-span of adults is very short, probably no more than a week. After mating the female lays up to 200 eggs on various host trees, including especially walnuts (*Juglans nigra*) but also hickories (*Carya* spp.), persimmon (*Diospyros virginiana*), and sweetgum (*Liquidambar stryaciflua*). At least in Missouri there are three broods each summer, over a flying period of from early April through August. Two types of egg are laid. One type hatches larvae that are fast-growing and able to hatch a second brood during the same season in which they are hatched, and a second type hatches larvae that pupate but remain dormant until the following breeding season to hatch and complete their life cycle. The tiny eggs hatch in about two weeks, and the larval stage lasts another three to four weeks in the fast-growing type. The large green larvae (up to almost three inches long) produce defensive clicking sounds with their mandibles when disturbed and can regurgitate a fluid that helps repel possible predators. At pupation the larva spins a silk thread around itself, and as the cocoon is formed the larva pulls a leaf around the silky structure, securing it to the leaf and concealing it. In the fast-growing type there is a three-week pupation period before the adult moth emerges.

## White-lined Sphinx
*Hyles lineata*
Adults of the white-lined sphinx can be easily identified by the prominent white stripes on their wing veins, a wide pink band on each hindwing, and a tan stripe on each forewing that extends from its base to its tip. This is one of the most common of the sphinx moths, and it can often be observed during late afternoons at flower gardens, where they seem to be especially attracted to white, tubular flowers. They are possibly a pollinator species of the rare western prairie fringed orchid. They are also attracted to evening porch lights. The white-lined sphinx has been reported from more than 30 Nebraska counties and is widespread across the United States; its range also extends to Central and South America, the West Indies, and even Hawaii. The moths lay their small green eggs singly or in clusters on various food plants. Their eggs hatch within several days, or if laid late in the summer may overwinter and hatch in the spring. The larvae are pale green with a series of nine or ten black-lined white oval spots extending along the sides of the abdomen, below which is a row of smaller yellow and black spots. All sphinx moth larvae have a sharp, rear-pointing "horn" near the tip of the abdomen, thus their colloquial name "hornworm." The horn perhaps serves as a visual defensive signal, and disturbed larvae of some species can produce a buzzing bee-like sound, or even aggressively strike if touched. The larvae are gluttonous, feeding continuously on their host plants. They eat leaves of plants from many host families, including those of the willow, rose, four-o-clock, and evening primrose families.

# 15 The Dragonflies and Damselflies of Spring Creek Prairie

This checklist is based on an in-house list of Spring Creek's dragonflies and damselflies. A more comprehensive list of Nebraska's species can be found in *The Nature of Nebraska* (Johnsgard, 2001b), and a complete online guide to Nebraska's Odonata is also available at http://museum.unl.edu/research/entomology/Odonata/anis.html. Many Nebraska species are also included in the excellent field guides by Manolis (2003) and Abbott (2005). Damselflies are much more difficult to field-identify than are dragonflies, but they are commonly seen and therefore some locally occurring species are included here. The taxa are arranged alphabetically, initially by genus and secondly by species. Measurements refer to the total distance from the front of the head to the abdominal tip. Plate ("Pl") references refer to photographs in *Dragonflies through Binoculars* (Dunkle, 2000); page numbers ("p.") refer to *Dragonflies and Damselflies of the West* (Paulson, 2009).

## Dragonflies (Anisoptera)

Like butterflies, dragonflies are a group of about 600 North American species that are instantly recognizable, but identifying individual species is another matter. It is easy to visually separate typical dragonflies from their subgroup, the damselflies: all dragonflies hold their wings horizontally outward at rest, whereas damselflies hold the wings vertically backward at rest. In dragonflies the forewings and hindwings differ in that their hindwings are broader at the base than are the forewings (Anisoptera means "unequal wings"); in most damselflies

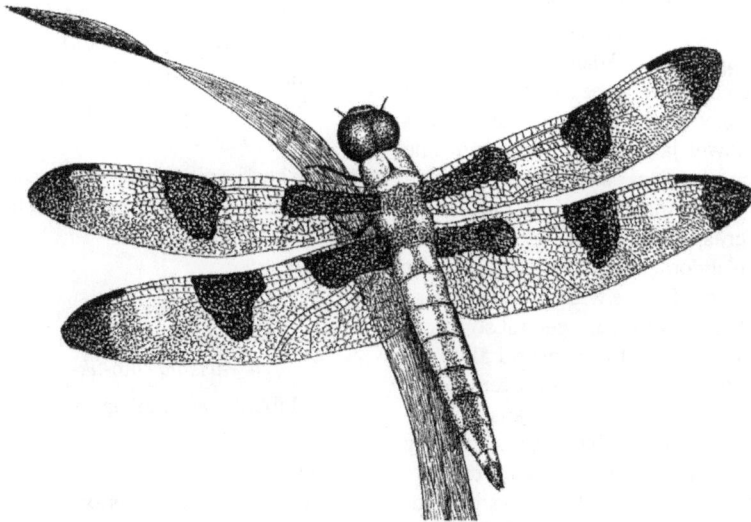

*Fig. 32. Twelve-spotted skimmer*

the forewings and hindwings are similar in size and shape (Zygoptera means yoke-shaped, or X-like, wings) and usually have stalked bases. Dragonflies are somewhat larger than damselflies; the approximately 50 species of Nebraska dragonflies range from 0.9 to 3.2 inches in length, whereas the 40-odd species of damselflies are 0.7 to 2.2 inches long.

Dragonflies and damselflies have minute antennae, but adults of both groups have extremely large, protruding eyes that probably allow for detailed 360-degree vision and effective visual hunting. They also have flexible "necks" that allow the head to be swiveled for better viewing. The brilliant and varied colors exhibited by adults of both groups indicate that their color vision is well developed. Differences in color patterns between the sexes and among different species suggest that species and sex recognition are both based on visual rather than olfactory cues. Sometimes color differences are related to age also, the young individuals being less colorful than adults. The sexes do not differ significantly in length, but females often have larger abdomens and usually differ in color, typically being browner overall.

Dragonflies and damselflies are effective predators, both as aquatic larvae and as winged adults. They have an extendable lower jaw (labium) with movable structures (palps) with hooks, teeth, and spiny hairs that can quickly close upon and grasp prey. In both groups the tip of the abdomen in males is adapted for grasping a female when mating. Males also have a secondary genital structure at the bottom of the second abdominal segment that serves for temporary sperm storage. Sperm is transferred to the female during a complicated mating behavior during which the male clings to the female's head or thorax as she bends her abdomen upward and forward to extract sperm from his abdominal storage

structure, as the two are flying about linked in a unique circular or "wheel" position.

All dragonflies are highly mobile, and some are seasonally migratory (or are long-distance wanderers), such as the wandering glider, spot-winged glider, common green darner, black saddlebags, and variegated meadowhawk. The migrations of the wandering glider (or globe-skimmer) evidently extend over thousands of miles across continents and oceans, and recent studies suggest that the species consists of a single worldwide genetic melting pot.

### Darners
*(large size; huge fused eyes; long, slender abdomen; strong fliers)*

Lance-tipped darner, *Aeshna constricta*. Widespread, ponds. 2.8–3.2 inches. Pl 4; p. 219.

Common green darner, *Anax junius*. Widespread, streams. 3.5–4.5 inches. Pl 1; p. 235.

### Emeralds and Basketails
*(emerald green eyes in contact dorsally; brown or black body)*

Common baskettail, *Epitheca cynosura*. Widespread. 1.6–1.9 inches. Pl 23; p. 360.

Prince baskettail, *Epitheca princeps*. Eastern Nebraska. 2.2–3.2 inches. Pl 23; p. 369.

### Clubtails
*(eyes separated; abdomen tip is variably club-like)*

Jade clubtail, *Arigomphus submedianus*. Eastern Nebraska. 2.1 inches. Pl 14; p. 249.

Plains clubtail, *Gomuhus externus*. Widespread. 2.1 inches. Pl 11; p. 266.

Sulfur-tipped clubtail, *Gomphus militaris*. Widespread. 2.0 inches. PI 7; p. 258.

### Skimmers
*(eyes in broad medial contact; wings often banded or tinted)*

Halloween pennant, *Celithemis euonina*. Widespread. 1.5 inches. PI 44; p. 419.

Eastern pondhawk, *Erythemis simplicicollis*. Widespread. 1.7 inches. PI 39; p. 441.

Widow skimmer, *Libellula luctuosa*. Widespread. 1.8 inches. PI 28; p. 390.

Common whitetail, *Libellula lydia*. Widespread. 1.7 inches. PI 28; p. 374.

Twelve-spotted skimmer, *Libellula obscurus*. Widespread. 2.0 inches. PI 29; p. 388.

Blue dasher, *Pachydiplax longipennis*. Widespread. 1–1.7 inches. PI 39; p. 483.

Wandering glider, *Pantala flavescens*. Widespread. 1.9 inches. PI 49; p. 513.

Spot-winged glider, *Pantala hymenaea*. Widespread. 1.9 inches. PI 40; p. 515.

Eastern amberwing, *Perithemis tenera*. Widespread. 0.9 inches. PI 38; p. 408.

Variegated meadowhawk, *Sympetrum costiferum*. Widespread. 1.5 inches. PI 35; p. 388.

Band-winged (Western) meadowhawk, *Sympetrum semicinctum*. Widespread. 1.3 inches. PI 36; p. 472.

Cherry-faced meadowhawk, *Sympetrum internum*. Widespread. 1.3 inches. PI 36; p. 469.

Autumn (Yellow-legged) meadowhawk, *Sympetrum vicinum*. Widespread. 1.3 inches. PI 35; p. 475.

Black saddlebags, *Tramea lacerata*. Widespread. 2.1 inches. PI 40; p. 511.

Red saddlebags, *Tramea onusta*. Southeast Nebraska. 1.8 inches. PI 41; p. 509.

## Damselflies (Zygoptera)

Damselflies have two pairs of similarly shaped wings that are held close together above the body at rest, or are only slightly divergent. Their eyes are widely separated. Damselflies also have legs with many long spines that are probably useful in aerial prey catching and have very thin abdomens as compared with dragonflies. Most of the Nebraska species are small and in the "bluets" group, the males of which are mostly blue, whereas the females may be similarly mostly blue or variably brown. In all the bluets the eyes are dark brown or black above and brightly colored below.

### Broad-winged Damselflies
*(Wings partly black or red; 1.5–2.2 inches)*

Ebony jewelwing, *Calopteryx maculata*. Woodland streams. 1.5–2.2 inches. p. 44.

American rubyspot, *Hetaerina americana*. Swift streams. 1.5–1.8 inches. p. 45.

### Pond Damselflies
*(Wings clear, stalked; body blue or brown, 0.7–1.7 inches)*

Red damsel, *Amphigrion abbrevistum*. Widespread. 1.0–1.2 inches. p. 130.

Paiute dancer, *Argia alberta*. Widespread. 0.7–0.9 inches. p. 148.

Blue-fronted dancer, *Argia apicalis*. Widespread. 1.3–1.7 inches. p. 143.

Springwater dancer, *Argia plana*. Widespread. 1.4–1.7 inches. p. 164.

Vivid dancer, *Argia vivida*. Vagrant. 1.3–1.6 inches. p. 166.

Rainbow bluet, *Enallagma antennatum*. Widespread. 1.2–1.4 inches. p. 95.

Azure bluet, *Enallagma aspersum*. Eastern Nebraska. 1.2–1.4 inches. p. 84.

Familiar bluet, *Enallagma civile*. Widespread. 1.2–1.7 inches. p. 81.

Skimming bluet, *Enallagma geminatum*. Widespread. 0.8–1.2 inches. p. 83.

Orange bluet, *Enallagma signatum*. Widespread. 1.2–1.6 inches. p. 99.

Eastern forktail, *Ischnura verticalis*. Widespread. 0.9–1.4 inches. p. 120.

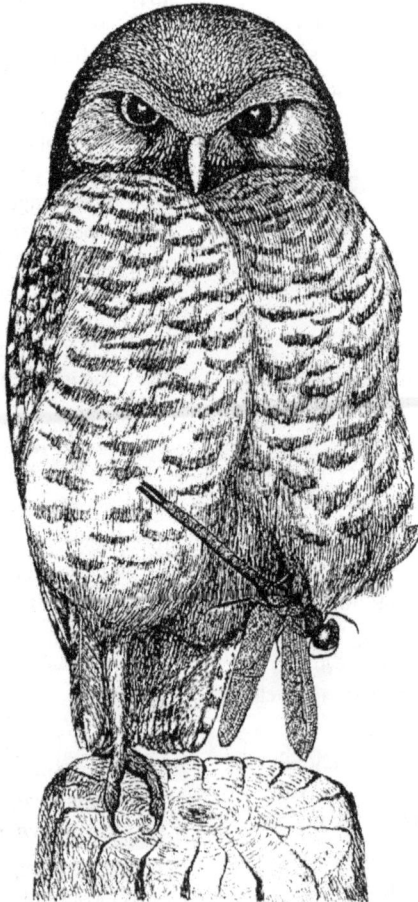

*Fig. 33. Common green darner and burrowing owl*

# 16  Profiles of Selected Dragonflies and Damselflies

**Green Darner**

*Anax junius*

The green darner is one of North America's largest and most common dragonflies. It is also one of the most widespread, ranging south from the northern states to Panama. It also occurs in the Caribbean and in China and Japan. Populations in the northern states migrate seasonally to the southern United States and Mexico. Males have emerald green eyes, head, and thorax, and a bright blue abdomen, whereas in females and immatures the abdomen is reddish or brown. In all individuals a dark dorsal line extends the entre length of the abdomen, and a black "bulls-eye" mark ringed with blue is on the top of the forehead.

The population that winters in Mexico is believed to breed upon their fall arrival, with the larvae developing there. Development of the larvae takes about a year, so the adults returning north and becoming the breeding population of the northern states are slightly more than a year old. Their offspring then undertake the fall migration southward. Darners are noted for strong territorial behavior. Males patrol their territory daily, and should misfortune befall a male and it disappears, he is likely to be immediately replaced by a successor (Sillsby, 2001). One apparently unique behavior of this species is that a male and female often fly in tandem while the female is laying her eggs on the water surface. One observer of green darners noted that migrating individuals in Canada would move from the west side of a leaf where they had settled in late afternoon to the east side of a leaf early the following morning, so they would gain warmth directly from the rising sun and be quickly able to begin their morning flights (Sillsby, 2001).

**Wandering Glider**

*Pantala flavescens*

This rather small dragonfly, the wandering glider, is also called the globe skimmer, which is an appropriate name considering its mobility. It has notably long and broad-based wings that allow it to glide and soar in the thermals, apparently eating aerial plankton in the upper atmosphere. It is a reddish-brown dragonfly with a yellow, tapered abdomen, a yellow face, and a small yellowish area at the base of the hindwings. Males develop an orange tinge on the dorsal side of the abdomen and have brown wingtips.

The wandering glider breeds on both sides of the Atlantic and Indian Oceans, and swarms of apparently migrating adults have been observed hundreds of miles at sea. Monsoon winds probably assist in these flights, carrying the insects to Africa, North and South America, Australia, and throughout Asia. They were one of the first species to appear in Bikini Atoll, South Pacific, after nuclear bomb testing was terminated, and for a time they colonized Easter Island, one of the most remote islands in the Pacific Ocean. Evidence indicates that millions of dragonflies, mostly wandering gliders, cross the Indian Ocean from southern India to eastern Africa. Leaving India in August, they arrive at the Maldives Islands in October, the Seychelles in November, and Mozambique in December. The return trip to India is made via the Maldives the following

April, completing an 11,000-mile round trip. It is believed that the migrants fly at a height of about 3,200 feet because winds at lower altitudes blow in the wrong direction.

## Bluets
*Enallagma* spp.
This general group of small damselflies is characterized by having blue abdomens. The genus *Enallagma* is a large assemblage with 34 North American species. In this group (subfamily Ischnurinae), a remarkable degree of female dimorphism and polymorphism occurs. In one form (morph) some females are the same color as males (andromorphs), but most differ in appearance (heteromorphs). The biological reasons for this diversity are still uncertain, but generally speaking it would appear that the andromorphs, being more colorful, are more likely to be attacked by birds and other predators, especially during oviposition. However, because of their female appearance, heteromorphs endure less harassment from males when they are ovipositing (Sillsby, 2001). It is believed that in this group many females mate only once during their lifetime, receiving enough sperm to last through their entire reproductive lives.

In one Australian species of *Ischnura* the males emerge about two weeks before the females. As soon as a female emerges she is seized by a male and mating occurs. After mating, the female flies high into the air, and air currents may carry her hundreds of miles. After four or five days her eggs will have developed, and egg-laying then occurs, possibly hundreds of miles from where mating had occurred nearly a week earlier (Sillsby, 2001).

# 17 The Grasshoppers and Katydids of Eastern Nebraska

Most of Nebraska's typical grasshoppers (Family Acrididae, the short-horned grasshoppers) have short antennae and hind legs adapted for jumping. More than 100 species of grasshoppers and katydids occur in Nebraska. Brust, Hoback, and Wright (2008) provide county distribution maps for all of Nebraska's short-horned grasshoppers, and more than 70 of Nebraska's more common species are listed by Johnsgard (2001b). Mechanical sounds (stridulation) are generated by males scraping chitinous skeletal structures over one another;

*Fig. 34. Courting Haldeman's grasshoppers (above) and a two-striped mermiria grasshopper (below)*

many of the grasshopper species strid-
ulate, usually to attract females. Typical
grasshoppers create these sounds by rubbing their hind legs against their forewings or their forewings against their
hindwings. Some grasshoppers also produce rasping noises by rubbing their
hind legs against their forelegs. Different sounds of the males have different
functions. The "calling song" is used to
attract females, and an aggressive, or
"fight," sound is used to establish dominance among males. The band-winged
grasshoppers produce noises by snapping their hind wings in flight, resulting
in crackling sounds called crepitation,
which might serve as an alarm function.
These sounds are apparently produced
when the rear wings are popped open.
Their rear wings also vary in color (red,
yellow, blue, gray, black) and pattern,
producing great variations among different species. Species with such colorful patterns exhibit them in conspicuous
wing-flashing displays.

Katydids (Family Tettigonidae) are
closely related to grasshoppers (and
sometimes are considered a subfamily of Acrididae). Most species are her-
bivorous, but some are predaceous.
Together with their great structural diversity, the males' conspicuous mating
"songs" (produced by rubbing their forewings together) are often species specific. Like other grasshoppers, katydids
(and crickets) produce their sounds by
mechanical means; a sharp edge at the
base of one of the forewings (scraper)
is rubbed over a file-like ridge on the
underside of the other front wing. Both
front wings have these structures, but in
katydids the right wing is usually the upper one. The file on the lower front wing
and corresponding scraper are usually
nonfunctional. In different species the
sounds ("songs") also differ, in the pulse
rate (each pulse is a single stroke of the
wings), the way the pulses are grouped,

and in the sound characteristics of the
pulses, such as clicking or buzzing. In
some katydids a "protest" sound is produced when threatened. Unlike grasshoppers, katydids sing only at night,
whereas crickets and typical grasshoppers sing during the day. Among some
crickets a courtship song is used prior to
mating. (In China, crickets are still often
kept in small cages so their songs can be
fully appreciated.)

No list of Spring Creek grasshopper
or cricket species yet exists; identification of many of the grasshoppers is very
difficult, even in the laboratory. Only
the conspicuous and colorful banded-
winged grasshoppers and the fairly easily recognized katydids are listed here.
However, *A Guide to the Tallgrass Prairies of Eastern Nebraska and Adjacent
States* (Johnsgard, 2008) lists 34 grasshoppers, 20 katydids, and 11 crickets of
eastern Nebraska's prairies. The excellent field guide by Capinera, Scott, and
Walker (2004) illustrates nearly all the
Nebraska species of grasshoppers, katydids, and crickets. In this list, the species described and illustrated in the Capinera field guide are identified by a #
symbol, with relevant text pages indicated. Taxa are arranged alphabetically,
initially by genus and secondarily by
species. Measurements refer to the distance from the front of head to the tip
of the wings (in long-winged forms), or
(in shorter winged species) to the tip of
the abdomen, but excluding the ovipositor in females.

**Typical Grasshoppers (Acrididae)**

***Band-winged Grasshoppers—***
*Subfamily Oedipodinae (hind wings
barred and often colorful, noisy in
flight; females larger than
males, maximum length
2.3 inches)*

Northwestern red-winged grasshopper, *Arphia pseudonietana*. 1.2–1.85 inches. Widespread, tall grasses. # p. 79 (wings bright orange-red with black tips)

Autumn yellow-winged grasshopper, *Arphia xanthoptera*. 1.2–1.8 inches. Eastern, grassy fields, woodland edges. # p. 80 (wings yellow to orange with black bands, large)

Northern green-striped grasshopper, *Chortophaga viridifasciata*. 1.1–1.5 inches. Widespread, short grasses. # p. 82 (wings white to yellow with gray bands, hind legs bluish)

Carolina grasshopper, *Dissoteira carolina*. 1.3–2.3 inches. Widespread and common, on open roadsides. # p. 88 (wings black with mottled yellow tips)

Dusky grasshopper, *Encoptolophus costalis*. 0.6–1.2 inches. Widespread, prairies and open grassland. # p. 90 (wings white with gray bands; blue hind legs)

Wrinkled grasshopper, *Hippiscus oceolote*. 1.1–2.1 inches. Widespread, pastures and weedy prairies. # p. 92 (wings yellow to pink with black bands)

Blue-legged grasshopper, *Metator pardulinus*. 1.0–1.8 inches. Widespread, many grassy habitats. # p. 95 (wings yellow to orange or rose with black bands; blue hind legs)

Haldeman's grasshopper, *Paradalophora haldemani*. 1.2–2.3 inches. Mainly western, weedy or sandy prairie. # p. 96 (wings bright orange to rose with black bands)

Kiowa rangeland grasshopper, *Trachyrhachys kiowa*. 0.8–1.2 inches. Widespread, bare gravelly ground. # p. 102 (wings yellow with dark gray bands)

## Katydids (Tettigonidae)

Katydids comprise a distinctive family of grasshoppers having very long antennae, long and bladelike ovipositor in females, and highly developed and species-typical sound production behavior (stridulation) by males. No Spring Creek Prairie species list yet exists; the following list is presumptive and based on Capinera et al. (2004). It includes all eastern and widespread Nebraska species, with text pages indicated by a # symbol. Species illustrated in *Insects in Kansas* (Salsbury and White, 2000) are indicated by asterisks. Taxa are arranged alphabetically, initially by genus and secondarily by species. Females are larger than males and often have extremely long ovipositors; head to abdomen-tip measurements exclude ovipositors.

*False Katydids—Subfamily Phaneropterinae (hindwings longer than forewings; wings variably leaflike; maximum length 2.2 inches)*

Oblong-winged katydid, *Amblycorypha oblongifolia*. 1.6–2.0 inches. East and central Nebraska, woodland understories. # p. 159.

Fork-tailed bush katydid, *Scudderia furcata*. 1.6–2.2 inches. Entire state, old fields and roadsides. # p. 163.

Texas bush katydid, *Scudderia texensis*. 1.6–2.0 inches. Entire state, old fields and roadsides. # p. 163.

*Cone-headed Katydids—Subfamily Copophorinae (head conelike; all wings and antennae very long; maximum 2.2 inches long)*

Sword-bearing conehead, *Neocono-cephalus ensiger*. 1.9–2.8 inches. Entire state, wet grassy areas. # p. 170.*

Nebraska conehead, *Neoconocephalus nebrascensis*. 1.9–2.8 inches. Eastern Nebraska, wet grassy areas. # p. 170.

Round-tipped conehead, *Neoconocephalus retusus*. 1.45–2.0 inches. Southeastern Nebraska, grassy or weedy areas. # p. 170.

Robust conehead, *Neoconocephalus robustus*. 2.3–3 inches. Entire state, moist upland prairies. # p. 170.

**Meadow Katydids**—Subfamily
Concocephalinae (forewings narrow;
very long antennae; maximum
1.6 inches long)

Slender meadow katydid, *Conocephalus fasciatus*. 0.7–1.0 inch. Entire state, common in many habitats. # p. 181.

Straight-lanced meadow katydid, *Conocephalus strictus*. 0.5–1.2 inches. Entire state, dry grasslands. # p. 182.

Gladiator meadow katydid, *Orchelimum gladiator*. 0.7–0.8 inch. Entire state, meadows. # p. 178.*

Common meadow katydid, *Orchelimum vulgare*. 1.0–1.6 inches. Entire state, pastures and fields. # p. 178.*

**Shield-backed Katydids**—Subfamily
Decticinae (forewings tiny, hidden;
short antennae; maximum
1.8 inches long)

Mormon cricket, *Anabrus simplex*. 1.1–1.8 inches. Entire state, scattered vegetation. # p. 187.

Haldeman's shieldback, *Pediodectes haldemanni*. 1.3–1.5 inches. Entire state, many habitats. # p. 186.

# 18 The Mantids and Walkingsticks of Eastern Nebraska

## Mantids (Mantidae)

Mantids are members of the large group of insects called Orthoptera ("straight-winged"), which also includes walking-sticks, grasshoppers, crickets, and cock-roaches. About 2,400 species are in the mantid family. Most of them are trop-ical, and some reach a body length of about 15 inches. In many species, the folded wings extend well beyond the tip of the abdomen. Some raise them above the body in a startling defensive posture, while making hissing sounds, when threatened. In some non-native genera (e.g., *Iris*, *Creobroter*), a colorful or somewhat eye-shaped wing pattern is then suddenly revealed. Mantids are carnivorous and have a distinctive "pray-ing" posture, adopted while waiting mo-tionless for prey. They are also distinc-tive in shape, with strong, spiny forelegs adapted for clutching and holding prey, an elongated neck, and a rather triangu-lar-shaped head that, almost uniquely among insects, can be pivoted about to search for prey, even behind them. Their eyes are large and oval in shape, sugges-tive of a classic science-fiction "alien" face. Mantids have stereoscopic vision, each eye having about 1,000 visual units (ommatidia). One area near the front of the eye (fovea) provides detailed vision, while the more peripheral areas detect movements. Many species of mantids are highly camouflaged by shape and color. They are often green, and their forewings sometimes closely resemble leaves. Their hindwings are transparent but are sometimes tinted or patterned with various colors. At least the longer-winged species can fly, and males are especially prone to fly at night, as they search for females, using the females' pheromones to track them.

Two of the known eastern Nebraska species are listed here. *Stagmomantis* is illustrated in *Insects in Kansas* (Sals-bury and White, 2000), and *Tenodera* is illustrated in Borror and White (1979). Measurements are from the front of the head to the tip of the wings when these are not elevated.

Carolina mantid, *Stagmomantis caro-lina*. 2–2.25 inches. Southeastern Nebraska, native.

Chinese mantid, *Tenodera aridifolia*. 3–4 inches. Eastern Nebraska, in-troduced.

## Walkingsticks (Phasmatidae)

Walkingsticks all closely resemble twigs and are highly inconspicuous, since they walk slowly and most species closely match their background. Their strange body forms have been a source of su-perstition; their family name derives from the Greek *phasma*, an apparition. Adults are often brownish, matching twig colors, whereas young individuals are grass-green, or when very young may be black, and mimic ants or scor-pions. Camouflage is highly developed in the insects resembling twigs or leaves, some of which even have moss-like out-growths on their bodes to enhance the mimicry. Most walkingsticks are wing-less, or nearly so, but some are winged and, like mantids, may raise and expose colorful wing patterns when threatened. Females are larger than males, and some species are surprisingly large; one in the southern states may reach seven inches in length. Some tropical species are extremely long, the largest being about 22 inches long, counting the extended legs, and weighing up to 65 grams. They have extremely long antennae; in some species these are as long as the body or even longer. Their legs are also very long and slender, and they have two types of pads, "toe pads" and "heel pads." Their toe pads are sticky and used for climbing on tilted or vertical substrates, whereas the heel pads are not sticky but have microscopic hairs that are able to provide traction on horizontal surfaces.

Walkingsticks are nearly unique among insects in being able to regenerate lost legs, at least to some degree. They eat primarily leafy vegetation, and when very abundant often damage trees. Unlike mantids, walkingsticks lay their eggs individually, and in most species they hatch in 20 to 30 days. Although essentially defenseless, when threatened walkingsticks can emit a foul-smelling gas whose components (terpenes) can cause eye damage at close range or burning sensations in the mouth. Some species also have sharp spines on their legs. They can make noise by rubbing parts of their wings together when threatened. Females lay from 100 to 1,200 eggs after mating, and many species reproduce without benefit of male involvement. Eggs laid in this "parthenogenetic" condition hatch as female clones of their mother. Both Nebraska species listed here are illustrated and described in *Insects in Kansas* (Salsbury and White, 2000). Measurements are from the front of the head to the abdominal tip.

Northern walkingstick, *Diapheromera femorata*. 2.95–3.75 inches. Eastern edge of Nebraska.

Prairie walkingstick. *Diapheromera velei*. 1.6–3.3". Common and widespread.

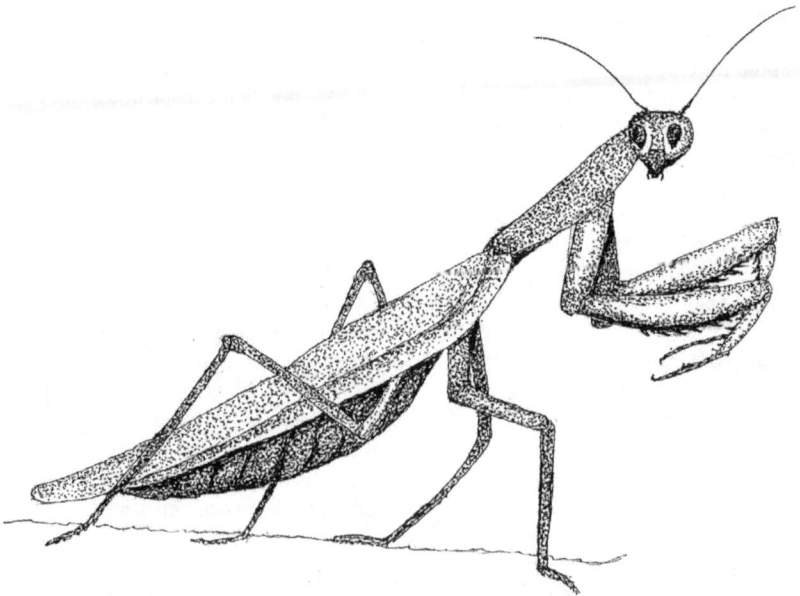

*Fig. 35. Chinese mantid, female*

## Profiles of Selected Mantids and Walkingsticks

### Carolina Mantid and Chinese Mantid
*Stagmomantis carolina* and *Tenodera aridifolia*

The range of Nebraska's only native mantid, the Carolina mantid, extends from Virginia to Florida and west to Mexico and California. The wings of this species do not reach the end of the abdomen, especially on females. The Chinese mantid, as its name suggests, is an import from Asia, after accidently being introduced by a nurseryman near Philadelphia in 1898. This is the largest mantid species in North America. Females are larger than males, ranging up to about 4.3 inches in length, and males average about 3.5 inches. Its long wings reach beyond the tip of the abdomen, and the forewings have a broad green stripe along their leading edge. Adults vary in color from green to brown, with oval eyes that match the body color, although in low light they appear black. The unworldly appearance of mantids may have been the basis for their Greek name *mantis*, meaning a prophet or seer. Not surprisingly, a host of European folklore and magical superstitions are associated with mantids, such as believing that rubbing the egg case, or a portion of one, on an infected or painful area will cure it immediately. No wonder the European species was named *Mantis religiosa*. This original "praying mantis" was introduced into the eastern United States in 1899, and is now established in many eastern states.

In one of his famous essays of the 1860s, Jean-Henri Fabre colorfully (if anthropomorphically) described the life of the European mantis (Stawell, 1921). He noted that the femur ("haunch") of the front legs have two rows of sharp teeth behind which are three spurs providing a "saw with two blades, behind which the leg (tibia) lies when folded back," providing a vice-like grip when closed. The tibia ("leg") has even more teeth than the femur, and is tipped with a sharp two-toed claw with "a double blade like a pruning knife." The prey is initially caught with the claw, then is drawn back between the saws as the vice-like device closes. Even animals as large as small mice and hummingbirds sometimes fall prey to such an ingenious killing device.

Females are famous for often killing and eating males immediately after having mated with them. Such behavior has often been attributed to helping put the female into prime laying condition, but the behavior is probably more common in captivity than under natural conditions, where the male's chances of escape would be better. After mating, females lay up to 400 eggs embedded in a frothy substance that is produced in the abdomen and expelled by the female. It hardens into a somewhat spherical structure. Fabre noted that as the female lays her eggs she simultaneously deposits this foamy material on the leaves or stems of vegetation. This matrix initially resembles soapsuds but soon solidifies into a cement-like hardness. In actuality, three types of materials are produced in three structural layers. The middle layer has two rows of slits ("doorways") through which the young mantids can escape after hatching. The eggs are arranged in layers, their ends containing the developing mantids' heads pointing toward the doors. Upon hatching, half of the young mantids exit through one "doorway" and the other half through the other (Stawell, 1921). Such amazing interior architecture of the entire structure could almost serve as a model for modern apartments. In a few species of mantids the mother guards the eggs until they hatch, but in many northern species the adults die before winter and the eggs do not hatch until spring.

## Northern Walkingstick and Prairie Walkingstick

*Diapheromera femorata* and *D. velei*

The prairie walkingstick is the common Nebraska walkingstick species. It is distributed from Minnesota and South Dakota south to Mexico. The larger northern walkingstick has a very restricted range in Nebraska but is very common in some states farther south. Walkingsticks are herbivores and eat the leaves of many native prairie plants. They are often inactive during the day, feeding all night. In places such as the Ouachita Mountains of Oklahoma and Arkansas, the northern walkingstick has sometimes defoliated and caused great damage to trees, especially oaks and other hardwoods. Female prairie walkingsticks are green and may be as large as 3.3 inches in length, whereas males are brown and may be as short as 1.6 inches. Both sexes are flightless and tend to drop to the ground when disturbed on vegetation. Their eyes are relatively small, unlike those of mantids, and it seems unlikely that their swaying behavior might enhance their judgment of distances by means of visual triangulation. Walkingsticks have tiny eyes with numerous facets, but light-sensitive photoreceptors (ocelli) are present only in males of the winged forms.

Like mantids, all walkingsticks are noted for their rhythmic slow-swaying behavior, which is of still uncertain function but has been attributed to improving their camouflage by mimicking the swaying of a leaf or twig. Another curious behavior of stick insects is their extremely long mating behavior. The smaller male might remain on the back of a female for days, weeks, or even longer. It has been suggested that the female might gain some protective benefits from having the male there to act as a sort of protective shield.

# 19 The Ecology of Nebraska's Tallgrass Prairies

The tallgrass prairie is one of the most romantic concepts of the American West. An imagined view of endless bison herds plodding through grasses so tall that they half obscured them from sight is a powerful image, and one that today must exist more in the realm of fancy than of fact. Quite probably most bison occurred on prairies of shorter stature, and the taller grasses that were present were likely soon clipped by the hungry migrants, but at least the vision of bison standing in belly-deep prairie is a most attractive one. One image that can still be realized is the sight of tallgrass prairie in full bloom from June through September, when dozens of prairie forbs vie for the attention of bees, butterflies, and moths as well as prairie-loving humans.

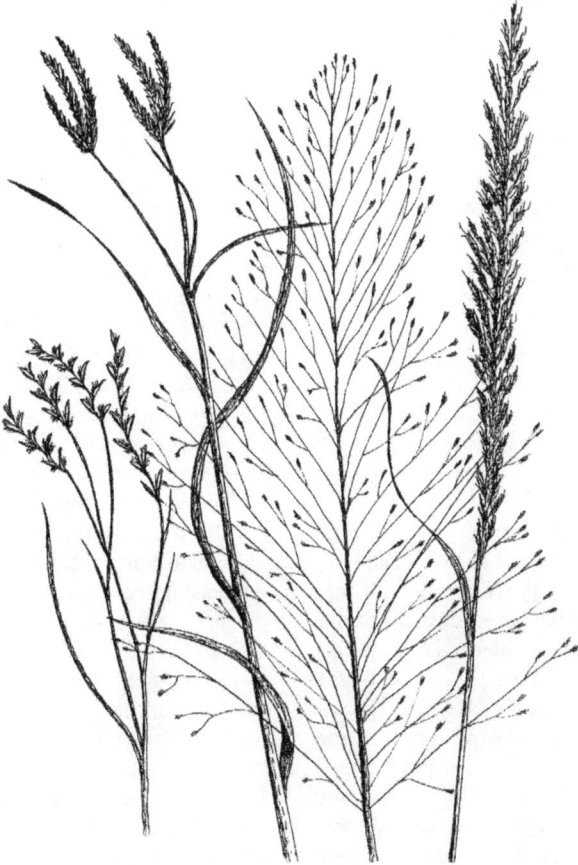

*Fig. 36. Prairie grasses (left to right):*
*little bluestem, big bluestem, switchgrass, and Indiangrass.*

Of all the grassland types in North America, the tallgrass prairie has been the most ravaged. One estimate of its original extent, based on a map published by A. W. Küchler (1964) was about 221,400 square miles, as compared with about 218,500 square miles for mixed-grass prairie and 237,500 square miles for shortgrass prairie. At least 95 percent of Nebraska's tallgrass prairie has now vanished; if the Sandhills prairies were classified as tallgrass prairie (they are usually considered to be mixed-grass prairie, but at times have been mapped as tallgrass prairie), they would certainly be the largest remaining remnant in all of North America. However, the species diversity of Sandhills prairie plants is much lower than in true tallgrass prairie, and plant density and floral diversity are both far less (Johnsgard, 1995).

One of the longest-studied of all tall-grass prairies is Nine-Mile Prairie near Lincoln. It has been studied more than half a century by John Weaver, T. L. Steiger, and other more recent botanists such as Robert B. Kaul and Steven B. Rolfsmeier. Nine-Mile Prairie had been reduced to only some 240 acres when it was purchased by the University of Nebraska Foundation in 1984, but it had totaled nearly 900 acres when Steiger (1930) was originally studying it.

Kaul and Rolfsmeier (1987) found 394 species of plants at Nine-Mile Prairie, of which 337 were native. There were 218 native perennials, 74 native annuals, 15 native biennials, 14 native shrubs, and 1 native tree. The composite family (Asteraceae) had the most species, 68, these being mostly summer- and fall-flowering species. The grass family (Poaceae) included 61 cool-season and warm-season species that collectively are in flower over the entire growing season. The legume family (Fabaceae) was the third largest in floral diversity with 23 species. Species endemic to the central continental region totaled 23 per-

cent, while 32 percent consisted of species with transcontinental biogeographic affinities. Those with central and eastern affinities composed 36 percent, and 8 percent had western and central North American distributional associations.

Similarly, Audubon's Spring Creek Prairie, of about 610 acres when acquired in 1998, was found by Kay Kottas (2000, 2001) to contain more than 349 species, of which 278 were native. Of these, 224 species were classified as upland species, 63 as wetland, and 62 as woodland adapted. Kottas's studies also revealed that Spring Creek Prairie had a higher overall plant species richness than Nine-Mile Prairie, although weedy and introduced species were more abundant at Spring Creek Prairie and contributed to this higher level of species diversity. The Spring Creek total included 74 species of Asteraceae, 69 species of Poaceae, and 29 species of Fabaceae. Among 15 species of warm-season grasses, big bluestem had the highest frequencies of occurrence in both prairies, followed at Spring Creek by little bluestem, side-oats grama, and Indiangrass, and at Nine-Mile Prairie by Indiangrass, little bluestem, and prairie dropseed. Other warm-season grasses present at both Spring Creek and Nine-Mile Prairie included purple lovegrass and oldfield three-awn.

The most abundant cool-season grasses at Spring Creek were (also in decreasing frequency) Kentucky bluegrass, smooth brome, Japanese brome, and Scribner's panicum. Smooth brome is a particularly troublesome invasive species in most Nebraska prairies, and Kentucky bluegrass is a similarly introduced and somewhat invasive species. Early spring burning (during late April, early May) is a common method of controlling cool-season grasses, and it also provides a sudden release of important mineral nutrients for use by warm-season species just prior to their period of rapid growth.

The pioneering research by T. L. Steiger (1930) at Nine-Mile Prairie revealed 345 species, of which 237 were identified as prairie species. He associated 70 of these with high prairie sites, 45 with low prairie, 77 with ravines, and 45 with wet meadows. The most commonly represented families were grasses with 38 species, composites with 46 species, legumes with 20 species, and sedges with 18 species. Although the grasses and sedges composed only 24 percent of the taxa, they represented 90 percent of the vegetation. Only ten of the 237 prairie species were annuals. The only trees present then were box elders (*Acer negundo*), cottonwoods (*Populus deltoides*), and three species of willows (*Salix*). Seven species of shrubs were also present.

Such famous Nebraska plant ecologists as John E. Weaver and Frederic Clements studied the prairies in eastern Nebraska for many decades. Their studies (e.g., Weaver, 1954, 1968; Weaver and Clements, 1954) established that about 200 species of upland forbs were typically present in tallgrass prairies, and that 75 of these were present in 90 percent of the prairies they examined. The most abundant and most consistently occurring upland forb is leadplant, which has a root system that can be up to more than 16 feet in length and has nitrogen-fixing root nodules. Many species of goldenrods are also usually present, with roots up to eight feet long.

Overall annual primary production of organic matter in tallgrass prairie averages about nearly 3,000 pounds per acre (300 grams per square meter). Likewise, the total underground parts of tallgrass prairie may contribute more than a ton of new organic matter per acre annually. Annual turnover (decomposition) rates for the above-ground parts of tallgrass prairie average about 80 percent, resulting in an average turnover period for the aboveground component of about 1.25 years, whereas turnover periods for underground biomass averages about three or four years. However, individual prairie perennial grasses and forbs have the potential for surviving for many decades.

As a result, prairie soils are constantly being refertilized by organic matter that has been produced during the past few growing seasons. The soils of tallgrass prairie are among the deepest and most productive for grain crops of any on earth. They represent the breakdown products of thousands of generations of annual productivity of grass and other herbaceous organic matter. Because of these organic materials and the clays usually present in prairie soils, such soils have excellent water-holding capabilities. In addition to the humus and related organic matter thus produced, many prairie legumes have nitrogen-fixing root bacteria that enrich and fertilize the soil to a depth of at least 15 feet. Earthworms and various vertebrate animals, such as gophers, make subterranean burrows that mix and aerate prairie soils, in the case of earthworms down to a depth of at least 13 feet.

## Dominant Plants of Wet Tallgrass Prairie

Although the plants of upland tallgrass prairie are impressive, those of the somewhat moister lowland prairie are even more so. In this situation big bluestem may compose 80 to 90 percent of the overall prairie vegetation, and together with little bluestem the two species represent at least 75 percent of all true prairie communities. Big bluestem is substantially taller than little bluestem and where both occur together the shorter species may be shaded out. On slopes and drier hilltops the smaller species has an advantage over the larger one. The roots of big bluestem extend about six to eight feet deep, and those of little bluestem are about five feet deep, so big bluestem has an advantage in moister sites. However, its roots tend to grow directly downward, whereas those of little bluestem and other bunchgrasses tend to spread widely, intercepting a much broader area than the aboveground parts of the plant.

Like many prairie perennials, both bluestem species are believed to be long-lived. Both species are warm-season grasses, and continue to grow through the summer. Big bluestem may rarely reach a height of 8 to 10 feet in some lowland sites by late summer, when it finally bursts into full flower. An additional 20 or more grass species are often present in lowland prairie. Weaver calculated that a strip of prairie sod four inches wide, eight inches deep and 100 inches long held a tangled network of roots having a total length of more than 20 miles! The total weight of prairie underground vegetation in the form of roots is likely to be as great as that of the aboveground parts, and much of this is recycled back into the soil on a yearly basis. In contrast, forests and woodlands store most of their productivity as woody aboveground parts, which recycle back into the soil only when the trees eventually die or are burned.

Besides big bluestem, Indiangrass, switchgrass, and in wetter sites, Canada wild rye and prairie cordgrass are the most important high-stature grasses of tallgrass prairie. All are at least five feet tall at maturity, and have root systems that extend down 8 to 12 feet for switchgrass, 7 to 8 feet for prairie cordgrass, and 5 to 6 feet for Indiangrass. All of these are warm-season grasses that are strongly rhizomatous. Two of the three are also continuous sod formers, but Indiangrass is a bunchier species, spreading mainly from lateral roots produced from late summer rhizomes that overwinter and provide for early

spring growth the following year. All these species reproduce mainly by rhizomes rather than from seed dispersal, and seed production estimates for big bluestem, Indiangrass, and switchgrass may average substantially less than that of little bluestem.

## Dominant Plants of Upland Tallgrass Prairie

The five dominant grasses of tallgrass prairie in upland situations, such as on hilltops and more sunny slopes, are plants of medium stature, and consist of little bluestem, needlegrass, prairie dropseed, Junegrass, and side-oats grama. All are bunchgrasses, and of all these perennial native grasses, little bluestem is easily the most important. It alone may compose 60 to 90 percent of the total vegetational cover, and on very favorable sites it may lose its bunching form and produce a continuous sod of interlocking roots. However, in most cases the major upland grasses occur in clumps spaced about a foot or more apart, with roots extending downward at least four to five feet.

Most of the important grasses of the tallgrass prairie are from three to six feet tall, with higher slopes having a greater proportion of mid-stature species. One of the few large and bushy shrubs to be of significance on the uplands is wild plum, although the smaller leadplant is widely distributed, and both prairie rose and New Jersey tea are likely to exist as scattered plants. There are also many summer- and fall-flowering composites, such as sunflowers, goldenrods, and asters, and the rare western prairie fringed orchid is sometimes found in moister locations. Many taller forbs are part of the low prairie flora. Taller shrubs, as are common in ravines, include wild plum, rough-leafed dogwood, and coralberry.

Forbs of the tallgrass prairie are numerous on the uplands. There, leadplant is usually the most important forb, although it has a woody base and might well be classified as a half-shrub. Other important half-shrubs include prairie rose and New Jersey tea. Other regular forb constituents of upland prairies are the prairie goldenrod, prairie flax, wild alfalfa, heath aster, bastard toadflax, and daisy fleabane.

The stiff sunflower is also one of the most widely distributed upland forbs, and it extends to many lowlands as well. Several other sunflowers, such as the saw-toothed sunflower, Maximilian's sunflower, compass-plant, and Jerusalem artichoke are also important prairie forbs, especially in moister situations. The Jerusalem artichoke is neither an artichoke nor native to Palestine; "Jerusalem" may have been a corruption of the Italian *girasole*, a sunflower. The "artichoke" portion of the name comes from the Arabic *al-krsufa*, which became the Greek word *alcachofe*—this perennial prairie plant has enlarged starchy tubers that can be eaten raw or cooked in various ways. Although tasty, the tuber

contains the carbohydrate inulin, which can't be digested but is metabolized by intestinal bacteria, often causing gastric distress (and giving rise to the unfortunate nickname fartichoke).

The broad-leaved scurf pea ("prairie turnip"), a once-common prairie legume, also has a tuber-like root that was historically an important food source for Native Americans but has become rather rare and is increasingly hard to find. The loss of prairie-pollinating insects through ignorant and unrestrained pesticide use is simply a recipe for massive insect genocide and probable long-term human suicide through losses of important human food plants.

*Of all things that live and grow upon this earth, grass is most important. . . . From the first oak openings of Ohio and Kentucky till it washed the foot of the Rockies, grass ocean filled the space under the sky. Steppe meadows, buffalo country, wide wilderness, where a man could call and call, but there was nothing to send back an echo.*

Donald Culross Peattie, *A Prairie Grove*

# 20 Profiles of Selected Tallgrass Prairie Grasses and Forbs

## Big Bluestem
*Andropogon gerardi*

Big bluestem is a tall warm-season grass, often reaching seven feet or more in height during the hot summer months. It finally bursts into blossom in September, when a person can walk, arms spread, through stands of big bluestem and come away with hands and shirt gilded by clouds of its golden pollen. By October it starts to shed its seed crop, which in natural stands might reach 100 pounds per acre, and much more in planted stands. By then its rather rank and drying foliage is not so attractive to large grazers, but earlier in the grow-

ing season it is a highly preferred food for most grazing mammals. *Andropogon* translates as "man's beard," a fair description of its flower head, which includes an equal mix of somewhat hairy and sessile but fertile spikelets and adjacent stalked but infertile ones. Sand bluestem (*Andropogon hallii*) is an extremely close relative of big bluestem that is more sand-adapted and arid-tolerant but is otherwise nearly identical, and sometimes the two forms hybridize where their ranges overlap in central Nebraska.

In addition to being the undisputed dominant of moist tallgrass prairie, big

*Fig. 37. Western prairie fringed orchid*

bluestem has an overall range extending east to the Atlantic coast, north in eastern Canada almost to James Bay, and south well into Mexico. Some other species of this genus have similar ranges. Little bluestem, sometimes placed in the same genus, is also a warm-season grass with a range similar to that of big bluestem, but it is shorter, more arid-adapted, and grows in clumps as a bunchgrass, rather than forming a continuous surface root layer. It is likely to be the dominant grass on prairie slopes and hilltops.

John Weaver once calculated that a square foot of big bluestem sod might contain about 55 linear feet of roots, and an acre of sod from the surface to a depth of only a few inches might hold about 400 miles of densely matted rhizomes. The strong roots of big bluestem have individual tensile strengths of 55 to 64 pounds, making prairie sod one of the strongest of natural organic substances. It is indeed strong enough to construct sod-built houses that have sometimes lasted a century or more in the face of Nebraska's inhospitable climate.

Weaver also calculated that big bluestem has a root system up to about three feet in diameter that can penetrate the soil to a depth of nearly seven feet. He determined that 43 percent of this grass's underground biomass is concentrated in the top 2.5 inches of soil, and 78 percent is present in the top six inches. The overall underground (root and rhizome) biomass of tallgrass prairies is usually two to four times greater than the above-ground biomass. The root component usually contributes about 30 percent of the annual primary production, or up to nearly 40 percent in the case of grazed prairie.

## Little Bluestem
*Schizachyrium scoparium*
This is the "shaggy" prairie grass of which Willa Cather wrote lovingly,

whose common name refers to a bluish cast that is present on the lower leaves and stem nodes of growing plants. However, by midsummer much of the entire visible plant starts to turn a rich Indian red, and by fall it is easy to recognize by its combination of bunch-like or "shaggy" shape and wonderful overall coppery red color, almost matching the colors of a autumnal prairie sunset. It and side-oats grama, whose equally distinctive florets that hang down one side of the plant stem like the feathers of a Lakota war lance (and was thus called "banner-waving-in-the-wind grass"), are two of the easily recognized and highly distinctive grasses of mixed-grass and tallgrass prairies.

Little bluestem is by far the most important plant of mixed-grass prairie, and it also extends eastward to share dominance with big bluestem on tallgrass prairie uplands and sunny slopes. It likewise penetrates the entire Sandhills region, and locally may even find opportunities for survival in moist depressions of shortgrass prairie. Like big bluestem it is a warm-season species, obtaining much of its growth in the warmest summer months, and sending out graceful feathery flowering stalks in early fall, typically in late September and October. Its abundant seeds are soon dropped, but the upright stems and leaves persist over the winter. In good years little bluestem may produce 200 or more pounds of seeds per acre, or at least as much as big bluestem. This compares with about 100 pounds of seed per acre produced by side-oats grama, and 100 to 180 pounds for blue grama. Cattle are not as fond of little bluestem for winter forage as are bison.

## Indiangrass
*Sorghastrum nutans*
Indiangrass probably won't ever have its name changed to "Native American grass" as a result of political correct-

ness, but I always associate its wonderful coppery fall color with the skin color of members that population, which I always envied among some of my mother's relatives, who still visibly carried dermal evidence of some of our family's ancestral genes. In Lakota lore, Indiangrass among all the tall prairie grasses is the one that is strongest and best able to stand up straight against adversities, in the manner of a Lakota warrior. The range of Indiangrass, like the historic range of the bison, extends over much of the North American Great Plains, reaching north to southern Manitoba. It also occurs east to Quebec and south to Florida, Texas, and central Mexico. Reaching six to seven feet tall, its flowering head is an upright spear-like cluster of light brown florets, painted with golden-colored stamens, which competes with big bluestem for being the finest attired of all the prairie grasses. It matures late, from September to November, when it is transformed by frost into a majestic copper-hued standard that stubbornly clings to its rich colors well into winter.

## Switchgrass
*Panicum virgatum*
Switchgrass is an important member of the tallgrass prairie community that grows best under rather moist and loamy soil conditions. In such situations it might easily reach five or six feet tall, but it thrives almost as well on sandy and clay soils. Switchgrass is also found in open woods, on gravel bars, and along stream banks. It primarily grows in clumps and in late summer is tipped with a finely textured crown of small pinkish florets that from a distance resemble smoke. It typically begins to blossom in July, and by fall its flower cluster turns beige, but eventually the florets dry and their numerous seeds fall, providing foods for sparrows and finches. Switchgrass grows north to the northern limits of tallgrass prairie in Manitoba,

east across southern Canada to Labrador, south into northern Mexico, and southwest to Baja California. It is hard to imagine that this wonderful grass ever served as a switch for punishing unruly schoolboys; it is more pleasant instead to think of it as a golden pennant designed to attract goldfinches.

## Western Prairie Fringed Orchid
*Plantathera praeclara*
The western prairie fringed orchid is a lovely, all too ephemeral, orchid that might remain hidden for years, suddenly to appear in full bloom for a week or so during late June or early July, then disappear as quickly and quietly as it had materialized, like some enchanting fairy spirit of childhood dreams. Thus, one must watch closely for it, especially in the wetter swales of tallgrass prairie. It has not yet been found at Spring Creek Prairie, but it is an indicator species of somewhat damp, species-rich virgin prairies, such as Wachiska Audubon's Dieken Prairie. A farmer-photographer friend told me of once haying in a prairie meadow and seeing its blooms just as the plant was about to be mowed down. Before he could stop the machine the flower had gone into the mower. Going back during following summers, he was never able to find the plant again. The plants often remain unseen for several years, in a dormant, subterranean state, nourished by micorrhizae. They might then suddenly exhibit a mass blooming, possibly stimulated by fire or by shifts in soil moisture that are associated with varied rainfall patterns.

There are many species of the genus *Plantathera*, most of which have whitish or greenish flowers and are pollinated by nocturnal or crepuscular moths. The white blossoms of the fringed orchid show up well under low-light conditions and no doubt help attract the moths. The enlarged and strongly fringed lower petal and sepals might also draw attention to the blossoms. Studies on the pollination biology by Charles Sheviak and Marlin Bowles (1986) have filled in the details for this species and a closely related but smaller one, the eastern prairie fringed orchid, which is fairly widespread in more eastern states.

Both species have blossoms that are creamy white to white, and in both the blossom fragrance is very sweet, intensifying after sunset. The blossoms of the western form are somewhat more creamy, and their fragrance more spicy, than in the eastern species. Their petal and sepal shapes also differ, and in the western species the blossom heads are shorter and denser, with fewer but larger individual blossoms. Both species are specifically adapted to pollination by sphinx moths, being nocturnally fragrant, deeply fringed, with extruded reproductive columns, and extremely long nectar-bearing spurs. Access to the spur is by a very limited entrance, and the pollinaria are situated in such a way that they will adhere either to the proboscis or eyes of the visiting moth. After the pollen-bearing structures have deposited their pollen on a moth, the columns rotate so they fully expose their stigmas, ready to receive pollen from the next moth that visits.

Sheviak and Bowles (1986) estimated that any pollinating moths of the western species must have a proboscis length between 35 and 45 millimeters, and must also have an across-the-eyes distance that approximates the distance between the sticky pollen-bearing structures (viscidia). At least five prairie-ranging sphinx moths seem to meet these requirements, all of which are native to Nebraska (the achemon, white-lined, wild cherry, laurel, and vashti sphinxes). Of these, the head measurements of the vashti sphinx does not quite "fit" the proper requirements, and it may only be a nectar thief, able to obtain nectar without carrying away pollen. The

same is possibly true of the wild cherry sphinx, and the laurel sphinx is rather rare. The remaining achemon and white-lined sphinxes both seem to qualify as possible pollinators.

Although it historically occurred all across eastern Nebraska, the current known distribution of the western prairie fringed orchid is limited to Lancaster, Seward, Otoe, Hall, and Cherry Counties. In 1989 the species was listed federally as a threatened species.

## Small White Lady's-slipper
### Cypripedium candidum

This beautiful little orchid once had a range similar to those of the eastern and western prairie fringed orchids combined. It extended west into eastern Nebraska (but not Lancaster County), and east to the southern New England states. It favors damp soil but full sunlight, often occurring in wetter meadows than where the prairie fringed orchid might also occur. Once very common in the wet meadows of eastern Nebraska, this orchid is now rare and is currently known only from four Nebraska counties. It is on the Nebraska list of threatened species.

This little lady's-slipper blooms fairly early, in May and June, or about the same time as the yellow lady's-slipper and before the white prairie fringed orchid. The blossoms may open before the leaves are fully unfurled, the flowers being mostly yellowish green except for the lower lip, which is glossy white with some flecks and narrow lines of purple. The conspicuous stamen-bearing structure is golden yellow with conspicuous crimson spots, the colors probably serving as insect attractants. There is usually only a single blossom per stem but sometimes two. However, the plants often grow in clumps, with stems up to 12 inches high, and with the long, oval leaves wrapping around the stem at their bases. The white slipper-shaped pouch is up to an inch in length, and the two lateral petal-like sepals are long, narrow, and rather twisted, and the dorsal hood is formed by a sepal that is also elongated and somewhat twisted.

The pollination ecology of this species is still little known but is probably much like that of a close European relative (*C. calceola*) that probably was separated from it during glacial periods. This species was one of the many orchids studied by Charles Darwin (1877). He discovered that orchid flowers of this pouch-like type act as "conical traps, with the edges inwards, like the traps which are sold to catch beetles and cockroaches." More recently, Davies (1998) described the pollination ecology of the pink lady's-slipper (*Cypripedium acaule*). Most members of this genus are pollinated by bees, but some are pollinated by flies (Li et al., 2012).

Insects are perhaps attracted by scent, or nocturnally by the conspicuous white color of the pouch. The crimson spots on the yellow stamen-bearing structure attract further attention, and the purple lines leading inward along the pouch perhaps act as false nectar-guides. The plant produces a variety of fragrances, some of which are similar to sex-attractant pheromones used by bees for attracting females. Insects that crawl into the pouch become trapped and can escape only by exiting through one of the two rear openings. In doing so, they must first brush the surface of the stigma, and later one of the anthers. This sequence prevents self-pollination of the flower. Most of the visitors are bees, especially solitary bees of various genera such as *Andrena*, a large and widespread group that dig nesting burrows in soil and are thus called "mining bees." Bumblebees can alight on the pouch but cannot enter, and some small bees and flies that can enter are too small to effect pollination.

## Compass-plant
*Silphium laciniatum*
Among the tallest of the prairie forbs is compass-plant, which might grow to about ten feet high with leaves that at times might be nearly two feet long. In younger plants especially, the leaves are twisted vertically, and the leaf axis is oriented almost perfectly north and south (thus the plant's common name). This trait allows the plant's photosynthetic cells to take advantage of early morning and late afternoon sunlight but does not expose the leaves to desiccation during midday hours when the sun is directly overhead. It has been said that early immigrants headed west took advantage of the compass-plant trait and judged directions from it, and could even find their way in the dark by feeling its leaves. A closely related species, the cup plant (*S. perfoliatum*), has opposite leaves united at their bases in such a way that a small cup-like structure is formed, which holds water after rains and probably likewise serves as an anti-desiccation adaptation. These plants are part of a group of forbs in the genus *Silphium* that have long been known for their reputed medicinal properties, and the Pawnee made an herbal drink from compass-plant for medicinal or spiritual reasons. It is unlikely that the immigrants' oxen gave any thought to the plant's possible health benefits as they munched on its tasty if rough surfaced leaves, but it is true that compass-plant is one of the forbs that quickly disappears when a prairie is subjected to heavy grazing.

*What a thousand acres of silphiums looked like when they tickled the bellies of buffalo is a question never again to be answered, and perhaps not even asked.*

Aldo Leopold, *A Sand County Almanac*

# 𝟚𝟙 The Plants of Spring Creek Prairie

This taxonomic list of about 260 species is based primarily on the prairie plants reported from Spring Creek Prairie (Kottas, 2000, 2001), exclusive of trees, aquatic species, and some woodland plants. It also includes a few prairie species mentioned in the text but not yet reported from Spring Creek Prairie. The list sequence is arranged alphabetically by descending taxonomic order: family, genus, species). Geographic terms (e.g., "Eastern third") refer to geographic distributions within the state of Nebraska. The number of species shown as occurring in Nebraska was based on Kaul, Sutherland, and Rolfsmeier's first edition (2006) of *The Flora of Nebraska*; they have since (2011) documented some additional species.

*Fig. 38. Grasshopper sparrow on annual sunflower*

*Status:*    I = Introduced
             N = Native

*Lifespan:*  A = Annual
             B = Biennial
             P = Perennial

*Habitat:*   D = Disturbed uplands
             R = Ravine
             U = Upland prairie
             W = Wetlands

*Flowering period:*
             Sp = Spring
             Su = Summer
             F = Fall
             Sp/Su = Spring & summer
             Sp/F = Spring to fall
             Su/F = Summer and fall

## Grasses and Sedges

### Grass Family—Poaceae
### (ca. 200 species in Nebraska)

Western wheatgrass, *Agropyron smithii*. Widespread. N, P, U, Su

Redtop bent, *Agrostis stolonifera*. Widespread, weedy. I, P, W, Su

Big bluestem, *Andropogon gerardii*. Mostly eastern. N, P, W, Su/F

Prairie three-awn, *Aristida oligantha*. Mostly eastern. N, A, D, Su/F

Red (Purple) three-awn, *Aristida purpurea longiseta*. Widespread. N, P, D, Su

Side-oats grama, *Bouteloua curtipendula*. Widespread. N, P, U, Su/F

Blue grama, *Bouteloua gracilis*. Widespread. N, P, U, Su/F

Hairy grama, *Bouteloua hirsuta*. Mostly western. N, P, U, Su

Smooth brome, *Bromus inermis*. Widespread, weedy. I, P, U, Sp/Su

Japanese brome, *Bromus japonicus*. Widespread, weedy. I, A, D, Sp/Su

Downy brome, *Bromus tectorum*. Widespread, weedy. I, A, D, Sp

Buffalo grass, *Buchloe dactyloides*. Mostly western. N, P, U, Su

Longspine sandbur, *Cenchrus longispinus*. Widespread, weedy. N, P, D, Su/F

Tumble windmillgrass, *Chloris verticillata*. Mostly southern, weedy. N, P, D, Su/F

Dichanthelium, *Dichanthelium (Panicum) acuminatum*. Widespread. N, P, D, Sp/F

Leiberg's dichanthelium, *Dichanthelium (Panicum) leibergii*. Eastern quarter. N, P, U, Sp/Su

Scribner's dichanthelium, *Dichanthelium (Panicum) oligosanthes scribnerianum*. Widespread. N, P, U, Sp/F

Large crabgrass, *Digitaria sanguinalis*. Widespread, weedy. I, A, D, Su/F

Barnyardgrass, *Echinochloa crus-galli*. Widespread, weedy. I, A, D, Su/F

Goosegrass, *Eleusine indica*. Southeastern, weedy. I, A, D, Su/F

Canada wild rye, *Elymus canadensis*. Widespread. N, P, U, Su/F

Stinkgrass, *Eragrostis cilianensis*. Widespread. I, A, D, Su/F

Purple lovegrass, *Eragrostis spectabilis*. Eastern three-quarters. N, P, D, Su/F

Sixweeks fescue, *Festuca (Vulpia) octoflora*. Widespread, weedy. N, A, D, Su/F

Fowl mannagrass, *Glyceria striata*. Widespread. N, P, W, Su/F

Foxtail barley, *Hordeum jubatum*. Widespread, weedy. N, P, R, Sp/Su

Little barley, *Hordeum pusillum*. Widespread, weedy. N, A, D, Sp/Su

Fall witchgrass, *Leptoloma (Digitaria) cognatum*. Southeast and east-central, sandy. N, P, U, Su/F

Prairie junegrass, *Koeleria pyramidata*. Widespread. N, P, U, Su

Plains muhly, *Muhlenbergia cuspidatum*. Widespread. N, P, U, Su/F

Marsh muhly, *Muhlenbergia racemosa*. Widespread, weedy. N, P, U, Su/F

Witchgrass, *Panicum capillare*. Widespread, weedy. N, A, D, F

Fall panicum, *Panicum dichotomiflorum*. Eastern half, weedy. N, A, W, F

Switchgrass, *Panicum virgatum*. Widespread, tallgrass prairies. N, P, U, F

Paspalum, *Paspalum setaceum*. Widespread. N, P, D, Sp/F

Reed canarygrass, *Phalaris arundinacea*. Widespread, weedy. N, P, W, Su

Timothy, *Phleum pratense*. Widespread. I, P, U, Su

Canada bluegrass, *Poa compressa*. Widespread. I, P, U, Sp/F

Kentucky bluegrass, *Poa pratensis*. Widespread, weedy. I, P, U, Sp/F

Tumblegrass, *Schedonnardus paniculatus*. Widespread, weedy. N, P, D, Sp/F

Little bluestem, *Schizachyrium* (*Andropogon*) *scoparium*. Widespread. N, P, U, Su/F

Yellow foxtail, *Setaria glauca*. Widespread, weedy. I, A, D, Su/F

Green foxtail, *Setaria viridis*. Widespread, weedy. I, A, D, Su/F

Indiangrass, *Sorghastrum nutans*. Widespread. N, P, U, Su/F

Johnsongrass, *Sorghum halepense*. Widespread, weedy. N, P, U, Su/F

Prairie cordgrass, *Spartina pectinata*. Widespread. N, P, W, Su/F

Prairie wedgetail grass, *Sphenopholis obtusata*. Widespread. N, P, U, Su

Tall dropseed, *Sporobolus asper*. Widespread. N, P, U, Su/F

Prairie dropseed, *Sporobolus heterolepis*. Widespread, native prairie. N, P, U, Su/F

Poverty dropseed, *Sporobolus vaginiflorus*. Eastern half, weedy. N, A, U, F

Needlegrass, *Stipa comata*. Mostly central and western. N, P, U, Sp/Su

Porcupine grass, *Stipa spartea*. Widespread, native prairies. N, P, U, Sp/Su

### Sedge Family—Cyperaceae
### (126 species in Nebraska)

Fescue sedge, *Carex brevior*. Widespread. N, P, RUW, Sp

Sun sedge, *Carex heliophila*. Widespread. N, P, U, Sp/Su

Meade's sedge, *Carex meadii*. Mostly eastern quarter. N, P, U, Sp/Su

Molesta sedge, *Carex molesta*. Mostly eastern quarter. N, P, U, Sp/Su

Fern flatsedge, *Cyperus lupulinus*. Central and southeastern. N, P, U, Su/F

## Broad-Leaved Herbs
## (Wildflowers and Other Forbs)

### Acanthus Family—Acanthaceae
### (3 species in Nebraska)

Fringeleaf ruellia, *Ruellia humilis*. Southeastern, tallgrass and open woods. N, P, U, Su

### Pigweed Family—Amaranthaceae
### (17 species in Nebraska)

Redroot pigweed, *Amaranthus retroflexus*. Widespread, weedy. N, A, D, Su/F

Common water hemp, *Amaranthus rudis*. Mostly eastern, weedy. N, P, W, Su/F

### Cashew Family—Anacardiaceae (5 species in Nebraska)

Poison ivy, *Toxicodendron* (*Rhus*) spp. Statewide; also a woody shrub or vine. N, P, U, Su

### Parsley Family—Apiaceae (= Umbelliferae) (31 species in Nebraska)

Spotted waterhemlock, *Cicuta maculata*. Widespread, near streams. I, B, R, Su

Wild parsley (Desert biscuitroot), *Lomatium foeniculaceum*. Eastern and Panhandle. N, P, U, Sp

Black snake-root, *Sanicula canadensis*. Widespread. N, B, R, Su

### Dogbane Family—Apocynaceae (4 species in Nebraska)

Hemp (Prairie) dogbane, *Apocynum cannabinum*. Widespread, weedy. N, P, W, Su

### Milkweed Family—Asclepiadaceae (17 species in Nebraska)

Swamp milkweed, *Asclepias incarnata*. Widespread. N, P, W, Su

Narrow-leaved milkweed, *Asclepias stenophylla*. Widespread. N, P, U, Su

Common milkweed, *Asclepias syriaca*. Mostly eastern. N, P, U, Su

Butterfly milkweed, *Asclepias tuberosa*. Eastern half. N, P, U, Su

Whorled milkweed, *Asclepias verticillata*. Widespread. N, P, U, Su

Green milkweed, *Asclepias viridiflora*. Widespread. N, P, U, Su

Spider milkweed, *Asclepias viridis*. Southeastern. N, P, U, Su

### Sunflower Family—Asteraceae (= Compositae) (243 species in Nebraska)

Common yarrow, *Achillea millefolium*. Widespread, weedy. N, P, U, Su

Common ragweed, *Ambrosia artemisiifolia*. Mostly eastern, weedy. N, A, D, Su/F

Western ragweed, *Ambrosia psilostachya*. Widespread. N, P, U, Su/F

Giant ragweed, *Ambrosia trifida*. Widespread, weedy. N, A, D, Su/F

Pussy-toes, *Antennaria neglecta*. Eastern half. N, P, U, Sp/Su

Common burdock, *Arctium minus*. Eastern half, weedy. I, B, D, Su/F

Silky wormwood, *Artemisia dracunculus*. Widespread, scattered. N, P, U, Su/F

Cudweed (White) sagewort, *Artemisia ludoviciana*. Widespread. N, P, U, Su/F

White (Heath) aster, *Aster* (*Symphyotrichum*) *ericoides*. Widespread. N, P, U, F

Aromatic aster, *Aster* (*Symphyotrichum*) *oblongifolius*. Eastern three-quarters. N, P, U, F

Silky aster, *Aster* (*Symphyotrichum*) *sericeus*. Eastern quarter. N, P, U, F

Panicled aster, *Aster simplex* (*lanceolatus*). Widespread. N, P, W, F

Nodding beggar-ticks, *Bidens cernua*. Widespread, weedy. N, A, W, F

Devil's beggar-ticks, *Bidens frondosa*. Widespread. N, A, W, F

Tall beggar-ticks, *Bidens vulgata*. Widespread, weedy. N, A, W, F

Tuberous Indian plantain, *Cacalia plantaginea* (*tuberosa*). Widespread. N, P, U, Su

Musk thistle, *Carduus nutans*. Widespread, weedy. I, B, D, Su

Tall thistle, *Cirsium altissimum*. Mostly eastern, weedy. N, P, U, Su/F

Flodman's thistle, *Cirsium flodmanii*. Widespread, weedy. N, P, UW, Su

Wavyleaf thistle, *Cirsium undulatum*. Widespread, weedy. N, P, U, Su

Horseweed, *Conyza canadensis*. Widespread, weedy. N, A, D, Su/F

Fetid marigold, *Dyssodia papposa*. Widespread. N, A, D, Su/F

Purple coneflower, *Echinacea angustifolia*. Widespread. N, P, U, Su

Daisy fleabane, *Erigeron strigosus*. Widespread in eastern half. N, A, U, Su/F

White snakeroot, *Eupatorium rugosum (Ageratina altissima)*. Eastern half, woods, weedy. N, P, DRUW, Su/F

Viscid euthamia, *Euthamia gymnospermoides*. Widespread. N, P, U, Su/F

Curly-top gumweed, *Grindelia squarrosa*. Widespread, weedy. N, P, D, Su/F

Common sunflower, *Helianthus annuus*. Widespread. N, A, D, Su/F

Sawtooth sunflower, *Helianthus grosseserratus*. Mostly eastern, damp to dry sites. N, P, U, F

Stiff sunflower, *Helianthus rigidus (pauciflorus)*. Eastern half. N, P, U, Su/F

Jerusalem artichoke, *Helianthus tuberosus*. Widespread, open moist sites. N, P, U, Su/F

False sunflower (Oxeye), *Heliopsis helianthoides*. Mostly eastern, open woods, weedy. N, P, U, Su/F

Hawkweed, *Hieracium longipilum*. Southeastern corner. N, P, U, Su

False boneset, *Kuhnia (Brickellia) eupatorioides*. Widespread. N, P, U, Su/F

Blue lettuce, *Lactuca oblongifolia (pulchella)*. Widespread, moist places. N, P, U, Su/F

Prickly lettuce, *Lactuca serriola*. Widespread, weedy. I, AB, D, Su/F

Rough gayfeather, *Liatris aspera*. Eastern half. N, P, U, Su/F

Dotted gayfeather, *Liatris punctata*. Widespread on loess, glacial till, and sand. N, P, U, Su/F

Skeletonweed, *Lygodesmia juncea*. Widespread. N, P, U, Su

False dandelion, *Microseris (Nothocalais) cuspidata*. Widespread. N, P, U, Sp

Prairie coneflower, *Ratibida columnifera*. Widespread. N, P, U, Su

Black-eyed susan, *Rudbeckia hirta*. Widespread, moist places. N, P, U, Su

Golden glow, *Rudbeckia laciniata*. Eastern half, moist places. N, P, R, Su

Lamb's-tongue groundsel, *Senecio integerrimus*. Scattered, mostly northern half. N, P, U, Sp

Prairie ragwort, *Senecio plattensis*. Widespread. N, P, U, Sp

Rosinweed, *Silphium integrifolium*. Eastern third. N, P, DU, Sp

Cup plant, *Silphium perfoliatum*. Eastern third, moist, low ground. N, P, R, Su

Canada goldenrod, *Solidago canadensis*. Widespread, dry to moist sites. N, P, U, F

Prairie goldenrod, *Solidago missouriensis*. Widespread. N, P, U, F

Gray goldenrod, *Solidago nemoralis*. Widespread, scattered. N, P, U, F

Rigid goldenrod, *Solidago rigida*. Widespread. N, P, U, F

Showy goldenrod, *Solidago speciosa*. Eastern third and northern quarter. N, P, U, F

Common dandelion, *Taraxacum officinale*. Widespread, weedy. I, P, D, Sp/F

Goat's beard (Western salsify). *Tragopogon dubius*. Widespread. I, B, D, Sp/Su

Meadow salsify, *Tragopogon pratensis*. Rare in a few south-central counties. I, B, D, Sp/Su

Baldwin's ironweed, *Vernonia baldwinii*. Southeastern two-thirds. N, P, U, Su

Cocklebur, *Xanthium strumarium*. Widespread. I, A, U, Su/F

### Borage Family—Boraginaceae
### (29 species in Nebraska)

Hoary puccoon, *Lithospermum canescens*. Eastern quarter. N, P, U, Sp/Su

Narrow-leaved puccoon, *Lithospermum incisum*. Widespread. N, P, U, Sp/Su

False gromwell, *Onosmodium molle*. Eastern two-thirds and northern half. N, P, U, Su

### Mustard Family—Brassicaceae
### (74 species in Nebraska)

Hoary cress, *Cardaria* (*Lepidium*) *draba*. Southeastern third and scattered, weedy. I, P, D, Sp

Whitlow grass, *Draba reptans*. Scattered. N, A, D, Sp

Pennycress, *Thlaspi arvense*. Widespread, weedy. I, A, D, Sp

### Caesalpinia Family—Caesalpiniaceae
### (6 species in Nebraska)

Partridge-pea, *Cassia chamaecrista*. Eastern Nebraska. N, A, D, Su/F

### Bellflower Family—Campanulaceae
### (12 species in Nebraska)

Blue lobelia, *Lobelia siphilitica*. Widespread. N, P, W, Su/F

Venus's looking glass, *Triodanis perfoliata*. Southeast quarter and scattered. N, A, D, Sp/Su

### Hemp Family—Cannabaceae
### (3 species in Nebraska)

Hemp, *Cannabis sativa*. Eastern half and scattered. I, A, D, Su/F

### Pink Family—Caryophyllaceae
### (35 species in Nebraska)

Sleepy catchfly, *Silene antirrhina*. Eastern quarter and scattered. N, A, D, Su

### Goosefoot Family—Chenopodiaceae
### (42 species in Nebraska)

Lamb's quarters, *Chenopodium berlandieri*. Widespread. N, A, D, Su/F

### Spiderwort Family—Commelinaceae
### (5 species in Nebraska)

Long-bracted spiderwort, *Tradescantia bracteata*. Eastern half and northern quarter, moist places. N, P, U, Su

### Morning-glory Family—Convolvulaceae (19 species in Nebraska)

Hedge bindweed, *Calystegia sepium*. Eastern half and scattered. N, P, D, Su/F

Field bindweed, *Convolvulus arvensis*. Eastern half and scattered, weedy. I, P, D, Sp/F

### Stonecrop Family—Crassulaceae
### (2 species in Nebraska)

Virginia stonecrop, *Penthorum sedoides*. Mostly eastern. N, P, W, Su/F

### Cucumber Family—Cucurbitaceae
### (4 species in Nebraska)

Bur cucumber, *Sicyos angulatus*. Southeastern. N, A, R, Su/F

### Horsetail Family—Equisetaceae
### (6 species in Nebraska)

Field horsetail, *Equisetum arvense*. Widespread. N, P, W, Sp

### Spurge Family—Euphorbiaceae
### (28 species in Nebraska)

Flowing spurge, *Euphorbia corollata*. Very eastern. N, P, U, Su/F

Toothed spurge, *Euphorbia dentata*. Very southeastern corner. N, A, D, Su/F

Snow-on-the-mountain, *Euphorbia marginata*. Widespread. N, A, D, Su/F

Eyebane, *Euphorbia nutans*. Eastern half. N, A, D, Su/F

### Bean Family—Fabaceae
### (= Leguminaceae) (ca. 100 species in Nebraska)

Canada milk-vetch, *Astragalus canadensis*. Scattered statewide, moist prairies, open woods. N, P, U, Sp

Ground-plum, *Astragalus crassicarpus*. Widespread, scattered. N, P, U, Sp

Platte River milk-vetch, *Astragalus plattensis*. Central and western. N, P, U, Sp/Su

White wild indigo, *Baptisia (Leucophaea) alba*. Southeastern corner. N, P, U, Sp

Plains wild indigo, *Baptisia (Leucophaea) bracteata*. Southeastern corner. N, P, U, Sp

Canada tickclover, *Desmodium canadense*. Scattered. N, P, DU, Su

Tick trefoil, *Desmodium illinoensis*. Eastern third. N, P, U, Su

Wild licorice, *Glycyrrhiza lepidota*. Widespread. N, P, W, Su

Bush-clover, *Lespedeza capitata*. Eastern half. N, P, U, Su/F

Alfalfa, *Medicago sativa*. Widespread (forage crop). I, P, D, Sp/F

Sweet-clover, *Melilotis albus*. Widespread (weed and forage crop). I, P, D, Sp/F

White prairie-clover, *Petalostemon (Dalea) candida*. Widespread, weedy. N, P, U, Su

Purple prairie-clover, *Petalostemon (Dalea) purpurea*. Widespread. N, P, U, Su

Silky prairie-clover, *Petalostemon (Da1ea) villosa*. Widespread. N, P, U, Su

Silver-leaf scurf-pea, *Psoralea (Pediomelum) argophylla*. Widespread. N, P, U, Su

Prairie-turnip, *Psoralea (Pediomelum) esculenta*. Widespread. N, P, U, Su

Wild alfalfa, *Psoralea (Pediomelum) tenuiflora*. Widespread. N, P, U, Su

Clovers, *Trifolium* spp. Scattered (forage plants). I, P, D, Sp/F

American vetch, *Vicia americana*. Widespread. N, P, U, Sp/Su

### Gentian Family—Gentianaceae
### (6 species in Nebraska)

Downy gentian, *Gentiana puberulenta*. Southeast and north-central. N, P, U, F

### Waterleaf Family—Hydrophyllaceae
### (4 species in Nebraska)

Waterpod, *Ellisia nyctelea*. Widespread. N, A, DRU, Sp

### Iris Family—Iridaceae
### (7 species in Nebraska)

White-eyed grass, *Sisyrinchium campestre*. Eastern third. N, P, U, Sp

### Mint Family—Lamiaceae (= Labitae)
### (49 species in Nebraska)

Rough false pennyroyal, *Hedeoma hispida*. Widespread. N, A, D, Sp/Su

Field mint, *Mentha arvensis*. Widespread. N, P, W, Su/F

Wild bergamot, *Monarda fistulosa*. Widespread. N, P, DU, Su

Catnip, *Nepeta cataria*. Widespread, weedy. I, P, DW, Su/F

Pitcher's (Blue) sage, *Salvia azurea* (*pitcheri*). Southeastern corner and scattered. N, P, DU, F

Leonard small skullcap, *Scutellaria parvula*. Eastern third, prairies and open woods. N, P, U, Su

American germander, *Teucrium canadense*. Widespread. N, P, RW, Su

### Lily Family—Liliaceae
### (32 species in Nebraska)

Wild onion, *Allium canadense*. Widespread, moist prairies, open woods. N, P, U, Su

Asparagus, *Asparagus officinalis*. Scattered statewide. I, P, DU, Sp

Solomon's seal, *Polygonatum biflorum*. Eastern half and northern third, moist deciduous woods. N, P, R, Su

### Flax Family—Linaceae
### (8 species in Nebraska)

Grooved (Prairie) flax, *Linum sulcatum*. Eastern half. N, A, U, Su

### Mallow Family—Malvaceae
### (15 species in Nebraska)

Velvet leaf, *Abutilon theophrasti*. Eastern half. I, A, D, Su/F

Plains poppy-mallow, *Callirhoe alcaeoides*. Southeastern quarter. N, P, U, Su

Purple poppy mallow, *Callirhoe involucrata*. Widespread. N, P, U, Sp/Su

### Four-o'clock Family—Nyctaginaceae
### (9 species in Nebraska)

Hairy four-o'clock, *Mirabilis hirsuta*. Scattered statewide. N, P, U, Su

Narrow-leaved four-o'clock, *Mirabilis linearis*. Scattered statewide. N, P, U, Su

Wild four-o'clock, *Mirabilis nyctaginea*. Widespread, weedy. N, P, DUW, Su

### Evening Primrose Family—
### Onagraceae
### (28 species in Nebraska)

Plains yellow evening primrose, *Calylophus serrulatus*. Widespread. N, P, U, Su

Willow-herb (fireweed), *Epilobium coloratum*. Scattered, eastern two-thirds. N, P, W, Su/F

Large-flowered gaura, *Gaura longiflora*. Southeastern. N, B, D, Su/F

Small-flowered gaura, *Gaura parviflora* (= *G. mollis*). Widespread, weedy. N, B, D, Su/F

Yellow evening primrose, *Oenothera villosa*. Widespread. N, B, D, Su/F

Common evening primrose, *Oenothera biennis*. Widespread. N, B, D, Su/F

### Orchid Family—Orchidaceae (19 species in Nebraska)

Nodding ladies' tresses, *Spiranthes cernua*. Scattered, tallgrass and mixed-grass prairies, marshes. N, P, UW, F

Early ladies' tresses, *Spiranthes vernalis*. Southeastern corner. N, P, U, Su

### Woodsorrel Family—Oxalidaceae (3 species in Nebraska)

Gray-green wood sorrel, *Oxalis dillenii*. Widespread, weedy. N, P, U, Sp/F

Yellow wood sorrel, *Oxalis stricta*. Eastern half and northern half, weedy. N, P, U, Su

Violet wood sorrel, *Oxalis violacea*. Eastern half, weedy. N, P, U, Sp

### Plantain Family—Plantaginaceae (10 species in Nebraska)

Woolly plantain (Indianwheat), *Plantago patagonica*. Widespread, weedy. N, A, D, Su

Blackseed plantain, *Plantago rugelii*. Eastern half, weedy. N, P, D, Su

### Phlox (Polemonium) Family— Polemoniaceae (16 species in Nebraska)

Slenderleaf collomia, *Collomia linearis*. Northern and Panhandle, prairies and woods. N, A, U, Su

Blue phlox, *Phlox divaricata*. Eastern fourth. N, P, U, Sp/Su

Prairie phlox, *Phlox pilosa*. Eastern fourth. N, P, U, Su

### Milkwort Family—Polygalaceae (4 species in Nebraska)

Whorled milkwort, *Polygala verticillata*. Northern quarter, south-central, and southeastern. N, P, D, Su/F

### Buckwheat Family—Polygonaceae (46 species in Nebraska)

Common knotweed, *Polygonum arenastrum*. Widespread. I, A, D, Su/F

Pink smartweed, *Polygonum bicorne*. Southern two-thirds and scattered. N, A, D, Su/F

Pennsylvania smartweed, *Polygonum pensylvanicum*. Widespread, weedy. N, A, DW, Su/F

Climbing false buckwheat, *Polygonum scandens*. Widespread, weedy. N, P, D, Su/F

Sheep sorrel, *Rumex acetosella*. Southeastern and north-central. I, P, U, Sp/Su

### Primrose Family—Primulaceae (10 species in Nebraska)

Fringed loosestrife, *Lysimachia ciliata*. Widespread. N, P, W, Su

### Buttercup Family—Ranunculaceae (39 species in Nebraska)

Meadow anemone, *Anemone canadensis*. Eastern third. N, P, W, Su

Candle anemone, *Anemone cylindrica*. Eastern, central, and northern. N, P, U, Su

Prairie larkspur, *Delphinium virescens*. Widespread. N, P, U, Su

Purple meadow rue, *Thalictrum dasycarpum*. Eastern, central, and northern, moist habitats. N, P, U, Su

### Rose Family—Rosaceae
### (55 species in Nebraska)

Wild strawberry, *Fragaria virginiana*. Eastern half, prairies and woods. N, P, U, Sp

Tall cinquefoil, *Potentilla (Drymocallis) arguta*. Eastern, central, and northern, prairies and open woods. N, P, U, Su

Sulfur cinquefoil, *Potentilla recta*. Eastern half and scattered northwest. N, P, DU, Sp/Su

### Madder Family—Rubiaceae
### (10 species in Nebraska)

Catchweed bedstraw, *Galium aparine*. Widespread. N, A, R, Su

Narrow-leafed bluets, *Hedyotis (Houstonia) nigricans*. Southern quarter, rocky places. N, P, U, Su

### Figwort Family—Scrophulariaceae
### (54 species in Nebraska)

Rough purple gerardia, *Agalinis aspera*. Eastern two-thirds. N, A, U, Su

Cobaea penstemon, *Penstemon cobaea*. Southeastern. N, P, U, Sp

Common mullein, *Verbascum thapsus*. Widespread, scattered. I, P, D, Su

Purslane speedwell, *Veronica peregrina*. Widespread. N, A, D, Sp/Su

### Nightshade Family—Solanaceae
### (19 species in Nebraska)

Clammy ground-cherry, *Physalis heterophylla*. Scattered eastern two-thirds. N, P, U, Su

Spearleaf ground-cherry, *Physalis longifolia*. Widespread. N, P, D, Sp

Virginia ground-cherry, *Physalis virginiana*. Eastern half. N, P, U, Sp

Black nightshade, *Solanum ptychathum*. Eastern half. N, A, D, Sp/F

Buffalobur, *Solanum rostratum*. Widespread, weedy; poisonous. N, A, D, Su/F

### Nettle Family—Urticaceae
### (5 species in Nebraska)

Pennsylvania pellitory, *Parietaria pensylvanica*. Widespread, shaded woods. N, A, R, Su

Stinging nettle, *Urtica dioica*. Widespread, moist woods, stream banks; skin irritant. N, P, W, Su

### Vervain Family—Verbenaceae
### (9 species in Nebraska)

Prostrate vervain, *Verbena bracteata*. Widespread. N, A, D, Sp/F

Blue vervain, *Verbena hastata*. Widespread, moist prairies and woods. N, P, UW, Su

Hoary (Woolly) vervain, *Verbena stricta*. Widespread. N, P, D, Su

White vervain, *Verbena urticifolia*. Eastern two-thirds and scattered. N, P, DU, Su/F

### Violet Family—Violaceae
### (17 species in Nebraska)

Prairie violet, *Viola pedatifida*. Eastern half, prairies and open woods. N, P, U, Sp

Blue prairie violet, *Viola pratincola*. Scattered statewide. N, P, U, Sp

Downy blue violet, *Viola sororia*. Scattered statewide. N, P, U, Sp

### Caltrop Family—Zygophyllaceae
### (1 species in Nebraska)

Puncture vine, *Tribulus terrestris*. Widespread, weedy. I, A, D, Su/F

## Shrubs and Woody Vines

All of these species of shrubs and woody vines are native perennials, except multiflora rose.

### Cashew Family—Anacardiaceae

Smooth sumac, *Rhus glabra*. Widespread.

Poison ivy, *Toxicodendron* (*Rhus*) *radicans*. Statewide (a woody vine or forb).

### Sunflower Family—Asteraceae (= Compositae)

Cudweed sagewort, *Artemisia ludoviciana*. Widespread.

### Honeysuckle Family—Caprifoliaceae

Elderberry, *Sambucus canadensis*. Mainly eastern Nebraska.

Western snowberry, *Symphoricarpos occidentalis*. Statewide.

Buckbrush (Coralberry). *Symphoricarpos orbiculatus*. Mostly southeast Nebraska.

### Staff-tree Family—Celastraceae

Climbing bittersweet, *Celastrus scandens*. Mainly eastern Nebraska; woody vine.

### Dogwood Family—Cornaceae

Rough-leaved dogwood, *Cornus drummondii*. Eastern Nebraska.

### Cypress Family—Cupressaceae

Eastern red cedar, *Juniperus virginiana*. Mainly eastern Nebraska; also a tree.

### Bean Family—Fabaceae

Leadplant, *Amorpha canescens*. Statewide, in better prairies.

False indigo, *Amorpha fruticosa*. Statewide.

### Currant Family—Grossulariaceae

Missouri gooseberry, *Ribes missouriensis*. Widespread, moist woods.

### Buckthorn Family—Rhamnaceae

New Jersey tea, *Ceanothus americanus*. Eastern Nebraska.

### Rose Family—Rosaceae

Wild plum, *Prunus americana*. Entire state.

Sand cherry, *Prunus besseyi*. Widespread, sandy habitats.

Eastern chokecherry, *Prunus virginiana*. Entire state.

Prairie wild rose, *Rosa arkansana*. Widespread, woodland edges.

Multiflora rose, *Rosa multiflora*. Introduced, uplands.

Black raspberry, *Rubus occidentalis*. Eastern Nebraska, riparian.

### Willow Family—Salicaceae

Peach-leaved willow, *Salix amygdaloides*. Widespread; shrub or tree.

Sand-bar willow, *Salix interior*. Widespread in wet sites; shrub or tree.

### Grape Family—Vitaceae

River-bank grape, *Vitis riparia*. Widespread; woody vine.

# 22 Checklist of Eastern Nebraska Tallgrass Prairie Plants

This alphabetic list of about 370 Nebraska tallgrass prairie plants excludes trees as well as woodland-adapted and aquatic species but otherwise includes nearly all of Spring Creek's grass, forb, and shrub flora. It also includes most of the Nine-Mile Prairie flora, which was initially surveyed and documented by Steiger (1930) and updated by Kaul and Rolfsmeier (1987). See Johnsgard (2008) for additional lists of Nebraska prairie biota, some prairie plant identification keys, and information on other regional prairies. That reference also includes the following plant list, organized alphabetically by Latin names rather than initially by English names.

☐ Alfalfa, *Medicago sativa*
☐ American bittersweet, *Celastrus scandens*
☐ American bugleweed, *Lycopus americana*
☐ American germander, *Teucrium canadense*
☐ Aromatic aster, *Aster oblongifolius*
☐ Asparagus, *Asparagus officinalis*
☐ Azure aster, *Aster oolentangiensis*

☐ Bald spikerush, *Eleocharis erythropoda*
☐ Barnyard grass, *Echinochloa crusgalli*
☐ Bastard toadflax, *Comandra umbellata*

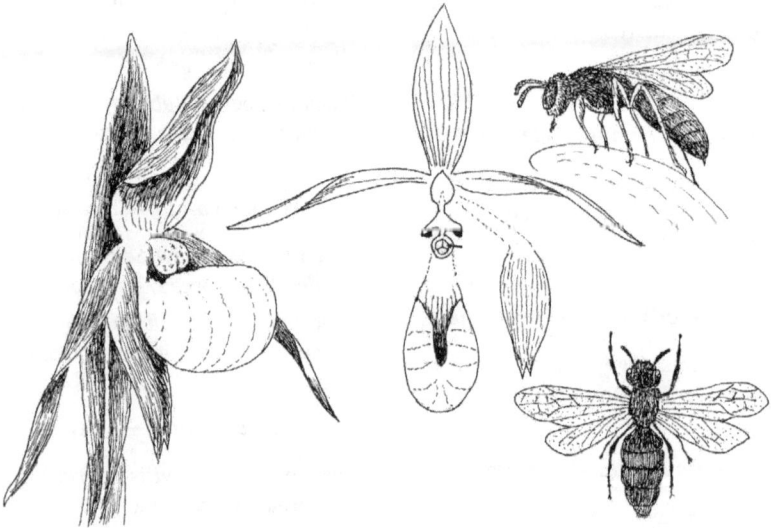

*Fig. 39. Small white lady's-slipper and Andrena bee*

- ☐ Bearded hawkweed, *Hieracium longipilum*
- ☐ Beggar-ticks, *Bidens vulgata*
- ☐ Big bluestem, *Andropogon gerardii*
- ☐ Black-eyed susan, *Rudbeckia hirta*
- ☐ Black medic, *Medicago lupulina*
- ☐ Black raspberry, *Rubus occidentalis*
- ☐ Black snakeroot, *Sanicula canadensis*
- ☐ Blazing star gayfeather, *Liatris squamosa*
- ☐ Blue grama, *Bouteloua gracilis*
- ☐ Blue lettuce, *Lactuca oblongifolia*
- ☐ Blue lobelia, *Lobelia siphilitica*
- ☐ Blue mustard, *Chorispora tenella*
- ☐ Blue prairie violet, *Viola pratincola*
- ☐ Blue vervain, *Verbena hastata*
- ☐ Bog yellow cress, *Rorippa palustris*
- ☐ Bottlebrush sedge, *Carex hysterica*
- ☐ Bristly greenbriar, *Smilax hispida*
- ☐ Broad-leaved cattail, *Typha latifolia*
- ☐ Buffalo bur, *Solanum rostratum*
- ☐ Buffalo grass, *Buchloe dactyloides*
- ☐ Bull thistle, *Cirsium vulgare*
- ☐ Bur cucumber, *Sicyos angulatus*
- ☐ Bushy wallflower, *Erysimum repandum*
- ☐ Butterfly milkweed, *Asclepias tuberosa*

- ☐ Canada bluegrass, *Poa compressa*
- ☐ Canada goldenrod, *Solidago canadensis*
- ☐ Canada milkvetch, *Astragalus canadensis*
- ☐ Canada wild rye, *Elymus canadensis*
- ☐ Candle anemone, *Anemone cylindrica*
- ☐ Carolina anemone, *Anemone caroliniana*
- ☐ Carpetweed, *Mollugo verticillata*
- ☐ Catchweed bedstraw, *Galium aparine*
- ☐ Catnip, *Nepeta cataria*
- ☐ Chenopodium, *Chenopodium strictum*
- ☐ Chickweed, *Stellaria pallida*

- ☐ Chicory, *Chichorium intybus*
- ☐ Chokecherry, *Prunus virginiana*
- ☐ Clammy ground-cherry, *Physalis heterophylla*
- ☐ Clearweed, *Pilea pumila*
- ☐ Climbing bittersweet, *Celastrus scandens*
- ☐ Climbing buckwheat, *Polygonum convolvulus*
- ☐ Cobaea penstemon, *Penstemon cobaea*
- ☐ Cocklebur, *Xanthium strumarium*
- ☐ Common arrowhead, *Sagittaria latifolia*
- ☐ Common burdock, *Arctium minus*
- ☐ Common evening primrose, *Oenothera biennis*
- ☐ Common knotweed, *Polygonum arenastrum*
- ☐ Common milkweed, *Asclepias syriaca*
- ☐ Common mullein, *Verbascum thapsus*
- ☐ Common ragweed, *Ambrosia artemisiifolia*
- ☐ Common sow thistle, *Sonchus asper*
- ☐ Common sunflower, *Helianthus annuus*
- ☐ Coralberry, *Symphoricarpos orbiculatus*
- ☐ Corn speedwell, *Veronica arvensis*
- ☐ Cudweed sagewort, *Artemisia ludoviciana*
- ☐ Cup plant, *Silphium perfoliatum*
- ☐ Curly dock, *Rumex crispus*
- ☐ Curly-top gumweed, *Grindelia squarrosa*
- ☐ Cursed crowfoot, *Ranunculus sceleratus*

- ☐ Daisy fleabane, *Erigeron strigosus*
- ☐ Dandelion, *Taraxacum officinale*
- ☐ Dark green rush, *Scirpus atrovirens*
- ☐ Deptford pink, *Dianthus armeria*
- ☐ Desert goosefoot, *Chenopodium pratericola*

☐ Devil's beggar-ticks, *Bidens frondosa*
☐ Dichanthelium, *Dichanthelium acuminatum*
☐ Ditch stonecrop, *Penthorum sedoides*
☐ Dotted gayfeather, *Liatris punctata*
☐ Downy blue violet, *Viola sororia*
☐ Downy brome, *Bromus tectorum*
☐ Downy gentian, *Gentiana puberulenta*

☐ Early ladies' tresses, *Spiranthes vernalis*
☐ Eastern black nightshade, *Solanum ptychanthum*
☐ Eastern chokecherry, *Prunus virginiana*
☐ Elderberry, *Sambucus canadensis*
☐ Emory's sedge, *Carex emoryi*
☐ English plantain, *Plantago lanceolata*
☐ Eyebane, *Euphorbia nutans*

☐ Fall panicum, *Panicum dichotomiflorum*
☐ Fall witchgrass, *Leptoloma cognatum*
☐ False boneset, *Kuhnia eupatorioides*
☐ False climbing buckwheat, *Polygonum scandens*
☐ False dandelion, *Microseris cuspidata*
☐ False gromwell, *Onosmodium molle*
☐ False indigo, *Amorpha fruticosa*
☐ False sunflower (ox-eye), *Heliopsis helianthoides*
☐ False toadflax, *Comandra umbellata*
☐ Fern flatsedge, *Cyperus lupulinus*
☐ Fetid marigold, *Dyssodia papposa*
☐ Field bindweed, *Convolvulus arvensis*
☐ Field horsetail, *Equisetum arvense*
☐ Field mint (wild mint), *Mentha arvensis*
☐ Field pennycress, *Thlaspi arvense*
☐ Field pussy-toes, *Antennaria neglecta*
☐ Field speedwell, *Veronica agrestis*

☐ Fire-on-the-mountain, *Euphorbia cyathophora*
☐ Fireweed, *Erechtites hieraciifolius*
☐ Flodman's thistle, *Cirsium flodmanii*
☐ Flowing spurge, *Euphorbia corollata*
☐ Four-o'clock, *Mirabilis nyctaginea*
☐ Fowl mannagrass, *Glyceria striata*
☐ Fox sedge, *Carex vulpinoidea*
☐ Foxtail barley, *Hordeum jubatum*
☐ Fringed loosestrife, *Lysimachia ciliata*

☐ Giant hyssop, *Agastache nepetoides*
☐ Giant ragweed, *Ambrosia trifida*
☐ Goat's beard, *Tragopogon dubius*
☐ Golden alexanders, *Zizia aurea*
☐ Golden glow, *Rudbeckia laciniata*
☐ Goosegrass, *Eleusine indica*
☐ Gray goldenrod, *Solidago nemoralis*
☐ Gray-green wood sorrel, *Oxalis dillenii*
☐ Gray sedge, *Carex grisea*
☐ Great Plains ladies'-tresses, *Spiranthes magnicamporum*
☐ Green foxtail, *Setaria viridis*
☐ Green milkweed, *Asclepias viridiflora*
☐ Grooved yellow flax, *Linum sulcatum*
☐ Ground plum, *Astragalus crassicarpus*

☐ Hairy crabgrass, *Digitaria sanguinalis*
☐ Hairy four-o'clock, *Mirabilis hirsuta*
☐ Hairy grama, *Bouteloua hirsuta*
☐ Hairy vetch, *Vicia villosa*
☐ Hairy wild rye, *Elymus villosus*
☐ Heath aster, *Aster ericoides*
☐ Heavy sedge, *Carex gravida*
☐ Hedge bindweed, *Calystegia sepium*
☐ Henbit, *Lamium amplexicaule*
☐ Hoary cress, *Cardaria draba*
☐ Hoary puccoon, *Lithospermum canescens*

- ☐ Hoary vervain, *Verbena stricta*
- ☐ Hog peanut, *Amphicarpaea bracteata*
- ☐ Hooded arrowhead, *Sagittaria calycina*
- ☐ Hornwort, *Ceratophyllum demersum*
- ☐ Horsenettle, *Solanum carolinense*
- ☐ Horseweed, *Conyza canadensis*

- ☐ Illinois tickclover, *Desmodium illinoense*
- ☐ Indiangrass, *Sorghastrum nutans*
- ☐ Indian hemp dogbane, *Apocynum cannabinum*
- ☐ Indian plantain, *Cacalia plantaginea*
- ☐ Inland rush, *Juncus interior*

- ☐ Japanese brome, *Bromus japonicus*
- ☐ Jerusalem artichoke, *Helianthus tuberosus*
- ☐ Jimson weed, *Datura stramonium*
- ☐ Johnny-jump-up, *Viola* (*bicolor*) *rafinesquii*
- ☐ Junegrass, *Koeleria pyramidata*

- ☐ Kentucky bluegrass, *Poa pratensis*
- ☐ Korean lespedeza, *Lespedeza* (*Kummerowia*) *stipulacea*

- ☐ Lady's thumb, *Polygonum persicaria*
- ☐ Lamb's-tongue groundsel, *Senecio integerimus*
- ☐ Large-flowered gaura, *Gaura longiflora*
- ☐ Leadplant, *Amorpha canescens*
- ☐ Leafy spurge, *Euphorbia esula*
- ☐ Leiberg's dichanthelium, *Dichanthelium leibergii*
- ☐ Leonard's small skullcap, *Scutellaria parvula*
- ☐ Little barley, *Hordeum pusillum*
- ☐ Little bluestem, *Schizachyrium scoparium*
- ☐ Long-bracted spiderwort, *Tradescantia bracteata*

- ☐ Lousewort, *Pedicularis canadensis*

- ☐ Maple-leaved goosefoot, *Chenopodium simplex*
- ☐ Marijuana (Hemp), *Cannabis sativa*
- ☐ Marsh muhly, *Muhlenbergia racemosa*
- ☐ Meadow salsify, *Tragopogon pratensis*
- ☐ Mead's sedge, *Carex meadii*
- ☐ Missouri goldenrod, *Solidago missouriensis*
- ☐ Missouri gooseberry, *Ribes missouriense*
- ☐ Missouri goosefoot, *Chenopodium missouriense*
- ☐ Moonseed, *Menispermum canadense*
- ☐ Motherwort, *Leonurus cardiaca*
- ☐ Multiflora rose, *Rosa multiflora*
- ☐ Musk thistle, *Carduus nutans*

- ☐ Narrow-flowered scurf pea, *Psoralium tenuiforum*
- ☐ Narrowleaf bluet, *Hedyotis nigricans*
- ☐ Narrow-leaved coneflower, *Echinacea angustifolia*
- ☐ Narrow-leaved four-o'clock, *Mirabilis linearis*
- ☐ Narrow-leaved milkweed, *Asclepias stenophylla*
- ☐ Narrow-leaved puccoon, *Lithospermum incisum*
- ☐ Needlegrass, *Stipa comata*
- ☐ New Jersey tea, *Ceanothus americanus*
- ☐ Nimblewill, *Muhlenbergia schreberi*
- ☐ Nodding beggar-ticks, *Bidens cernuus*
- ☐ Nodding fescue, *Festuca obtusa*
- ☐ Nodding ladies' tresses, *Spiranthes cernua*
- ☐ Norwegian cinquefoil, *Potentilla norvegica*

☐ Old-field balsam, *Gnaphalium obtusifolium*

☐ Oldfield three-awn, *Agrostis oligantha*

☐ Orchardgrass, *Dactylis glomerata*

☐ Pale dock, *Rumex altissimus*

☐ Pale-seeded plantain, *Plantago virginica*

☐ Panicled aster, *Aster simplex (lanceolatus)*

☐ Partridge pea, *Cassia chamaecrista*

☐ Pennsylvania pellitory, *Parietaria pensylvanica*

☐ Pennsylvania smartweed, *Polygonum pensylvanicum*

☐ Peppergrass, *Lepidium densiflorum*

☐ Philadelphia fleabane, *Erigeron philadelphicus*

☐ Pineapple weed, *Matricaria matricarioides*

☐ Pink smartweed, *Polygonum bicorne*

☐ Pitcher sage, *Salvia azurea*

☐ Pitseed goosefoot, *Chenopodium berlandieri*

☐ Plains muhly, *Muhlenbergia cuspidatum*

☐ Plains poppy-mallow, *Callirhoe alcaeoides*

☐ Plains wild indigo, *Baptisia bracteata*

☐ Plains yellow primrose, *Calylophus serrulatus*

☐ Platte River milk-vetch, *Astragalus plattensis*

☐ Poison hemlock, *Conium maculatum*

☐ Poison ivy, *Toxicodendron radicans*

☐ Pokeweed, *Phytolacca americana*

☐ Porcupine-grass, *Stipa spartea*

☐ Poverty grass, *Sporobolus vaginiflorus*

☐ Prairie coreopsis, *Coreopsis tinctoria*

☐ Prairie cordgrass, *Spartina pectinata*

☐ Prairie dropseed, *Sporobolus heterolepis*

☐ Prairie flax, *Linum sulcatum*

☐ Prairie goldenrod, *Solidago missouriensis*

☐ Prairie larkspur, *Delphinium virescens*

☐ Prairie ragwort, *Senecio plattensis*

☐ Prairie sandreed, *Calamovilfa longifolia*

☐ Prairie sedge, *Carex bicknellii*

☐ Prairie three-awn, *Aristida oligantha*

☐ Prairie trefoil, *Lotus purshianus*

☐ Prairie turnip, *Psoralea esculenta*

☐ Prairie violet, *Viola pedatifida*

☐ Prairie wedgegrass, *Sphenopholis obtusata*

☐ Prairie wild rose, *Rosa arkansana*

☐ Prickly lettuce, *Lactuca serriola*

☐ Prostrate vervain, *Verbena bracteata*

☐ Puncture vine, *Tribulus terrrestris*

☐ Purple-leaved willow-herb, *Epilobium coloratum*

☐ Purple lovegrass, *Eragrostis spectabilis*

☐ Purple meadow rue, *Thalictrum dasycarpum*

☐ Purple poppy mallow, *Callirhoe involucrata*

☐ Purple prairie clover, *Dalea purpurea*

☐ Red clover, *Trifolium pratense*

☐ Red-root flatsedge, *Cyperus erythrorhizos*

☐ Redtop, *Agrostis stolonifera*

☐ Reed canary grass, *Phalaris arundinacea*

☐ Rhombic copper leaf, *Acalypha rhomboidea*

☐ Rice cutgrass, *Leersia oryzoides*

☐ River-bank grape, *Vitis riparia*

☐ Rosin-weed, *Silphium integrifolium*

☐ Rough false pennyroyal, *Hedeoma hispidus*

☐ Rough gayfeather, *Liatris aspera*

☐ Rough-leaved dogwood, *Cornus drummondi*

- [ ] Rough pigweed, *Amaranthus retroflexus*
- [ ] Round-headed bush clover, *Lespedeza capitata*
- [ ] Round-leaved mallow, *Malva rotundifolia*
- [ ] Rugel's plantain, *Plantago rugelii*
- [ ] Russian olive, *Elaeagnus angustifolia*
- [ ] Rusty flatsedge, *Cyperus odoratus*

- [ ] Sandbur, *Cenchrus longispinus*
- [ ] Sand cherry, *Prunus besseyi*
- [ ] Sand paspalum, *Paspalum setaceum*
- [ ] Sawbeak sedge, *Carex stipata*
- [ ] Sawtooth sunflower, *Helianthus grosseserratus*
- [ ] Schweinitz's sedge, *Cyperus schweinitzii*
- [ ] Scribner's dichanthelium, *Dichanthelium oligosanthes*
- [ ] Sedge, *Carex molesta*
- [ ] Sensitive brier, *Schrankia nuttallii*
- [ ] Shattercane, *Sorghum bicolor*
- [ ] Sheep sorrel, *Rumex acetosella*
- [ ] Shepherd's purse, *Capsella bursa-pastoris*
- [ ] Short-beaked sedge, *Carex brevior*
- [ ] Showy goldenrod, *Solidago speciosa*
- [ ] Showy tick trefoil, *Desmodium canadense*
- [ ] Sideoats grama, *Bouteloua curtipendula*
- [ ] Silky aster, *Aster sericeus*
- [ ] Silky wormwood, *Artemisia dracunculus*
- [ ] Silver-leaf scurf-pea, *Psoralea argophylla*
- [ ] Skeletonweed, *Lygodesmia juncea*
- [ ] Sleepy catchfly, *Silene antirrhina*
- [ ] Slender gerardia, *Agalinis tenuifolia*
- [ ] Small duckweed, *Lemna minor*
- [ ] Smallflower buttercup, *Ranunculus arbortivus*
- [ ] Small-flowered gaura, *Gaura parviflora*

- [ ] Small white lady's-slipper, *Cypripedium cadidum*
- [ ] Smartweed dodder, *Cuscuta polygonorum*
- [ ] Smooth brome, *Bromus inermis*
- [ ] Smooth scouring rush, *Equisetum laevigatum*
- [ ] Smooth sumac, *Rhus glabra*
- [ ] Snow-on-the-mountain, *Euphorbia marginata*
- [ ] Soft-stem bulrush, *Schoenoplectus tabernaemontani*
- [ ] Solomon's seal, *Polygonatum biflorum*
- [ ] Spearleaf groundcherry, *Physalis longifolia*
- [ ] Spider milkweed, *Asclepias viridis*
- [ ] Standley's goosefoot, *Chenopodium standleyanum*
- [ ] Stickseed, *Hackelia virginiana*
- [ ] Stiff goldenrod, *Solidago rigida*
- [ ] Stiff sunflower, *Helianthus rigidus*
- [ ] Stinkgrass, *Eragrostis cilianensis*
- [ ] St. John's-wort, *Hypericum perforatum*
- [ ] Straw-colored flatsedge, *Cyperus strigosus*
- [ ] Sulphur cinquefoil, *Potentilla recta*
- [ ] Sun sedge, *Carex heliophila*
- [ ] Swamp milkweed, *Asclepias incarnata*
- [ ] Switchgrass, *Panicum virgatum*

- [ ] Tall dropseed, *Sporobolus asper*
- [ ] Tall fescue, *Festuca arundinacea*
- [ ] Tall gayfeather, *Liatris pycnostachia*
- [ ] Tall hedge mustard, *Sisymbrium loeselli*
- [ ] Tall nettle, *Urtica dioica*
- [ ] Tall thistle, *Cirsium altissimum*
- [ ] Tansy mustard, *Descurainia pinnata*
- [ ] Three-square bulrush, *Schoenoplectus pungens*
- [ ] Tick trefoil, *Desmodium illinoense*
- [ ] Timothy, *Phleum pratense*

- ☐ Toothcup, *Ammannia robusta*
- ☐ Toothed spurge, *Euphorbia dentata*
- ☐ Tumblegrass, *Schedonnardus paniculatus*

- ☐ Velvet-leaf, *Abutilon theophrasti*
- ☐ Venice mallow, *Hibiscus trionum*
- ☐ Venus' looking glass, *Triodanis perfoliata*
- ☐ Violet wood sorrel, *Oxalis violacea*
- ☐ Virginia creeper, *Parthenocissus quinquefolia*
- ☐ Virginia groundcherry, *Physalis virginiana*
- ☐ Virginia wild rye, *Elymus virginicus*
- ☐ Viscid euthamia, *Euthamia gymnospermoides*

- ☐ Water hemp, *Amaranthus rudis*
- ☐ Watermeal, *Wolffla columbiana*
- ☐ Waterpod, *Ellisia nyctelea*
- ☐ Water smartweed, *Polygonum punctatum*
- ☐ Wavyleaf thistle, *Cirsium undulatum*
- ☐ Western ironweed, *Vernonia baldwinii*
- ☐ Western ragweed, *Ambrosia psilostachya*
- ☐ Western rock jasmine, *Androsace occidentalis*
- ☐ Western snowberry, *Symphoricarpos occidentalis*
- ☐ Western wallflower, *Erysimum asperum*
- ☐ Western wheatgrass, *Agropyron smithii*
- ☐ White avens, *Geum canadense*
- ☐ White clover, *Trifolium repens*
- ☐ White-eyed grass, *Sisyrinchium campestre*
- ☐ Whitegrass, *Leersia virginica*
- ☐ White lettuce, *Prenanthes aspera*

- ☐ White mulberry, *Morus alba*
- ☐ White prairie clover, *Dalea candida*
- ☐ White snakeroot, *Eupatorium rugosum*
- ☐ White (Cudweed) sagewort, *Artemisia ludoviciana*
- ☐ White sweet clover, *Melilotus albus*
- ☐ White vervain, *Verbena urticifolia*
- ☐ White whitlow-wort, *Draba reptans*
- ☐ White wild indigo, *Baptisia lactea*
- ☐ Whorled milkweed, *Asclepias verticillata*
- ☐ Whorled milkwort, *Polygala verticillata*
- ☐ Wild alfalfa, *Psoralea tenuiflora*
- ☐ Wild bean, *Strophostyles leiosperma*
- ☐ Wild bergamot, *Monarda fistulosa*
- ☐ Wild lettuce, *Lactuca canadensis*
- ☐ Wild licorice, *Glycyrrhiza lepidota*
- ☐ Wild onion, *Allium canadense*
- ☐ Wild parsley, *Lomatium foeniculaceum*
- ☐ Wild petunia, *Ruellia humilis*
- ☐ Wild plum, *Prunus americana*
- ☐ Wild strawberry, *Fragaria virginiana*
- ☐ Willow-leaved lettuce, *Lactuca saligna*
- ☐ Windmill grass, *Chloris verticillata*
- ☐ Winter cress, *Barbarea vulgaris*
- ☐ Wirestem muhly, *Muhlenbergia frondosa*
- ☐ Woodbine, *Parthenocissus vitacea*
- ☐ Wood sedge, *Carex blanda*
- ☐ Woolly plantain, *Plantago patagonica*

- ☐ Yarrow, *Achillea millefolium*
- ☐ Yellow foxtail, *Setaria glauca*
- ☐ Yellow sweet clover, *Melilotus officinalis*
- ☐ Yellow wood sorrel, *Oxalis stricta*
- ☐ Yerba de tajo, *Eclipta prostrata*

# 28 Selected Public-Access Regional Tallgrass Prairies

Anyone wanting to observe tallgrass prairie birds and native plants should consider visiting some of the following tallgrass prairies owned by the Wachiska chapter of the National Audubon Society, Lincoln, Nebraska. About 70 more public-access grasslands are described in *A Guide to the Tallgrass Prairies of Eastern Nebraska and Adjacent States* (Johnsgard, 2008), and many prairies in other states are described in *A Naturalist's Guide to the Great Plains* (Johnsgard, 2018). Visitors should respect both the plant and animal life of these fragile sites, which often support rare, threatened, or endangered species. Precautions for avoiding sunburn, wood ticks, mosquitoes, and poison ivy should also be considered, and having a supply of water along is crucial.

*Fig. 40. Greater prairie-chicken, male*

An excellent book for the identification of Nebraska wildflowers is John Farrar's *Field Guide to Wildflowers of Nebraska and the Great Plains*, which is organized by flower color for ease of use. The most recent edition (2011) includes 281 species. Two other more regional guides organized by flower color are Ladd's *Tallgrass Prairie Wildflowers* (1995), which describes 285 species, and Denison's *Missouri Wildflowers* (2008), which illustrates 297 plants. There is also Johnson and Larson's (1999) book *Grassland Plants of South Dakota and the Northern Great Plains*, which includes 289 species. It is organized taxonomically and is especially useful for grass and sedge identification. Several other regional guides are listed in the references section. Latin names are shown here for a few species; others are identified elsewhere.

### Public-Access Prairies Described in This Book

### Spring Creek Prairie Audubon Center

*Owned by the National Audubon Society. 850 acres in Lancaster County, including about 600 acres of native prairie and 250 acres of restored prairie, woodlands, and wetlands.*

Drive three miles south on W 98th St. from the western edge of Denton; the entrance gate is on the east side of the road. Native tallgrass prairie uplands, some small wetlands, including a spring, and riparian deciduous (oak, ash, cottonwood) woodland. An old wagon trail crosses the property, and walking trails over the high hills offer wonderful panoramic views. The bird list includes about 230 species, and there is a prairie-chicken lek. The vascular plant list totals more than 370 species. The modern straw-bale nature center has restrooms, water, educational exhibits, a gift shop, and artwork. Open Monday–Friday 9 a.m. to 5 p.m.; call or check the website for weekend hours and special events. Admission fee $4 adults, $3 seniors and children. Memberships are available. Free admission on Tuesdays. Phone 402-797-2301. http://springcreek.audubon.org/

### Nine-Mile Prairie

*Owned by the University of Nebraska Foundation. 240 acres of native upland prairie in Lancaster County.*

Located nine miles northwest of the official center of Lincoln (9th and "O" Streets). Drive four miles west on W "O" St. in Lincoln, then four miles north on NW 48th St. to West Fletcher Ave., then one mile west. Park in the small parking lot on the north side; don't drive through the gate and enter the restricted developed area with its many buildings. Walk south about 100 yards from the parking lot through a metal gate to the prairie; beyond another approximately 100 yards a walking trail heads west along a metal fence and the prairie edge. Open all hours daily. Free admission. The total plant list exceeds 390 species. http://snr.unl.edu/aboutus/where/fieldsites/ninemileprairie.aspx

### Public-Access Prairies Owned by Wachiska Audubon Society

*Wachiska Audubon*
*4547 Calvert St., Suite 10*
*Lincoln, NE 68506*
*(402-486-4846)*

All Wachiska-owned prairies have public access. The following text is reproduced with the permission of Wachiska Audubon, with minor editorial changes and additions. See the organization's prairies website at https://www.wachiskaaudubon.org/mentoring-programs.

## Berg Prairie East and Berg Prairie West

*9.9 and 11.7 acres, respectively, Johnson County, near Talmage, Nebraska*

Berg Prairie has two parts. The East prairie is on bottomland of the Little Nemaha River, and the land surrounding it is in the Wetland Reserve Program. It has quite a lot of compass-plant and other wet-prairie species.To reach the Berg prairies, go eight miles south from the south edge of Syracuse to Talmage Road, then nine miles east to Hwy. 67. The East prairie lies on the east side of Hwy. 67, a half-mile east and a half-mile south of Talmage. The West prairie is a drier prairie in the interior of the section across the road and is less accessible. This is an excellent prairie, with great native plant diversity. To reach it, go north of the East prairie to the first intersection and then west a half-mile, where you will see the prairie sign. Wachiska has right-of-way access along the east side of the fence to reach this prairie, which requires a one-fourth mile walk.

## Clarence and Ruth Fertig Tallgrass Prairie

*42 acres, Colfax County, near Richland, Nebraska*

Fertig Prairie is located between Schuyler and Columbus near the Platte River, about an hour and 45 minutes from Lincoln. This is a wet prairie, which should remain actively flowering all summer. Wild onions are abundant in early June, along with purple poppy mallow, prairie larkspur, sensitive brier, and prairie wild rose. To find the prairie, proceed west five miles from Schuyler on Hwy. 30. At the junction of Hwy. 30 and Co. Rd. 6 turn south at the large 4-H sign with the green four-leaf clover logo. Follow Co. Rd. 6 (minimum maintenance gravel) south for 2.5 miles to the prairie. If the road is wet, it might be best to park at the intersection of Co. Rd. 6 and Rd. D, and then walk about 150 yards to the prairie, which is on the right side near the end of the dead-end road, and has a metal gate (but no permanent sign yet as of spring 2018). During very wet weather, it would be advisable to take Co. Rd. 7 south and then go west the final mile on Co. Rd. D.

## Henry Dieken Tallgrass Prairie

*14 acres, Otoe County, southwest of Unadilla, Nebraska*

Good wildflower displays can be seen from spring to fall at this priaire. Visit in late May to mid-June for prairie phlox (on east-facing slope) and narrow-flowered scurf pea. Mid-June to early July is the best time for finding New Jersey tea, black-eyed susan, and leadplant. Prairie fringed orchids have been seen in low swales in early July during at least two relatively moist summers. Rough gay-feather and compass-plant are in bloom during August. Downy gentian and nodding ladies' tresses are present during September and October but may be hard to find among thick, higher vegetation. To reach Dieken Prairie, drive 1.5 miles south on Rd. 20 from the northwest corner of Unadilla and the NE Hwy. 2/Rd. 20 intersection. Proceed to the intersection with Rd. I, and then go west 0.75 mile on Rd. I. The prairie is located on the south side of Rd. I in the northwest corner of the northwest quarter-section.

## Clyde and Thelma Gewacke Prairie

*11 acres, Fillmore County, west-north-west of Ohiowa or east-northeast of Strang, Nebraska*

This prairie has an excellent stand of needle-and-thread grass. As a historical note, it was named in honor of original property owners Clyde and Thelma Gewacke, the parents of Margaret (Gewacke) Nichols. The prior property owner was Margaret Nichols. To reach Gewacke Prairie, drive 2.5 miles west from Ohiowa Spur Road. Alternatively, drive five miles east from the NE Hwy. 74/US 81 intersection (at the northwest edge of Strang) to the intersection with Rd. 18, and then go 0.5 mile north on Rd. 18. The prairie is on the west side of road, in the southeast corner of the northeast quarter-section.

### Elmer Klapka Farm and Prairies
*400 acres total, Pawnee County, southeast of Table Rock, Nebraska*
Native prairies are situated in three separate locations on this large, one-time farm property (a map is needed for precise locations). They include 25 acres that are in excellent condition at three sites, 111 acres that are in fair/good condition at three other sites, and 32 acres that are in poor condition at one site. The balance of the property is in various stages of prairie restoration. From the east edge of Table Rock and the intersection of NE Hwy. 4 and Pawnee St., proceed east-southeast on Pawnee St. until it intersects with Ave. 626 (about 0.25 mile); then go south about 2.25 miles on Ave. 626 to the intersection with Rd. 712. About 120 prairie acres are located immediately southeast of this intersection. Approximately 240 acres are located about 0.5 mile east on the north side of Rd. 712, and about 40 acres are one mile east, on the south side of Rd. 712.

### Ivan A. and Ivan F. Lamb Tallgrass Prairie
*6.3 acres, Johnson County, southwest of Sterling, Nebraska*
This small hillside prairie is packed with wildflowers. Especially prominent are leadplant, purple prairie coneflower, plains (common) evening primrose, and purple prairie clover. The best time to visit for seeing flowers is late June to early July. Drive two miles west from Sterling's west end on NE Hwy. 41/43

to the intersection with Ave. 608 (a Lutheran church is on north side of the road at this intersection). Then drive two miles south across the BNSF railroad tracks to the intersection with Rd. 732. The prairie is on the southwest corner of the intersection, in the northeast corner of the northeast quarter-section.

### Knott Tallgrass Prairie
*21 acres, Saunders County, northeast of Yutan, Nebraska*
A lowland Platte River floodplain prairie, this is a good location for seeing bobolinks in late May and June. A rare plant, the wood betony or lousewort (*Pedicularis canadensis*), blooms in May. Tall (thick-spike) gayfeather and compassplant bloom from mid-July to early August. During summer, mosquitoes may be very evident in this wet prairie! This prairie has previously been known as both Yutan Prairie and Storm Prairie. To find it, drive one mile north on Rd. 5 from the Rd. 5/NE Hwy. 92 intersection at the southeastern edge of Yutan to Rd. N. Then go one mile east on Rd. N to the intersection with Rd. 4. Finally, proceed 0.5 mile north on Rd. 4 (which is often wet and rutted). *Note:* These are distinctly unimproved gravel roads. If Rd. N is not passable, proceed one more mile north on Rd. 5 from Yutan to Rd. O, then one mile east on Rd. O to the end of Rd. O at the intersection with Rd. 4. Finally, turn right and drive 0.5 mile south on Rd. 4 to the prairie. The prairie is on the east side of the road and is bordered by an evergreen grove on the north and an NRCS Conservation Easement prairie to the south.

### Wildcat Creek Tallgrass Prairie
*30.5 acres, Gage County, south-southwest of Virginia, Nebraska*
This expansive and photogenic prairie has one of the best displays of orange-red-blossomed butterfly milkweeds any-

where. They bloom mainly in June and July, rarely into September. Get to Wildcat Prairie by driving seven miles south on Liberty Rd. (S 162nd Rd.) from the northwest corner of Virginia and NE Hwy. 4/Liberty Rd. At the stop-sign intersection on B-Line Rd. proceed one mile west on B-Line Rd. to the intersection with S 148th Rd. Then drive one mile north on S 148th Rd. to E. Osage Rd., and finally go 0.5 mile west on E. Osage Rd. The prairie is located on the north side of E. Osage Rd., which is an accurately described "minimum maintenance road."

## Prairies Open to Limited Public Access

### Fricke Cemetery Prairie

*5 acres, Richardson County, north-northeast of Falls City, Nebraska*

This small cemetery prairie is thick with wildflowers from mid-June to early July. Golden alexanders, prairie phlox, purple coneflower, New Jersey tea, and leadplant are common. In mid-October, the brilliant scarlet tints of smooth sumac and the cerulean-hued azure asters (*Aster oolentangiensis*) make a wonderful visual combination. No advance notice is needed before visiting Fricke Cemetery. To find it, drive five miles north on US Hwy. 73 from the north end of Falls City to its intersection with Rd. 712. Alternatively, drive nine miles east of Verdon on US 73/Rd. 712 to the intersection of US Hwy. 73 and Rd. 712. Then go 3.5 miles east on Rd. 712 to its intersection with Ave. 655, and finally go 0.5 mile north on Ave. 655. The prairie is on the west side of Ave. 655, in the northeast corner of the quarter-section and west of a farmstead across Ave. 655. Please park on the road and walk in; also, close the gate after entering or leaving the prairie!

### Linwood Cemetery Prairie

*6 acres, Butler County, south of Linwood, Nebraska*

Linwood Cemetery is a bluff-top prairie with a splendid view of the Platte valley. It has many of the typical upland prairie wildflowers, including skeletonweed, common evening primrose, silver-leaf scurf pea, and leadplant. The best time for seeing flowers is from mid-June to mid-July. No advance notice is needed before visiting. Drive one mile south on Rd. X from the southeast corner of Linwood to the end of Rd. X. The prairie is located immediately south of the Rd. X terminus.

### Bentzinger Prairie

*13.5 acres, Johnson County, between Syracuse and Tecumseh, Nebraska*

Bentzinger Prairie is privately owned. No advance notice is needed before visiting, but be sure to treat Louis and Grace Bentzinger's prairie with care. This prairie has a fine array of both upland and some wet prairie wildflower species. Narrow-flowered scurf pea and plains yellow primrose start appearing in late May. By mid-June, butterfly milkweed, prairie phlox, and New Jersey tea are also blooming. In early July, leadplant, purple prairie clover, prairie coreopsis, and blazing star gayfeather are blooming in the upland areas. Tall white wild indigo blooms in June, and tall gayfeather blossoms during July in the wetter areas. To reach Bentzinger Prairie, drive 10.5 miles south on NE Hwy. 50 from the NE Hwy. 50/NE Hwy. 2 intersection at Syracuse. Or, drive 9.5 miles north on NE 50 from the NE Hwy. 50/US 136 intersection north of Tecumseh. The prairie is on the west side of NE Hwy. 50 in the northeast corner of the northeast quarter-section. It is bordered by NE Hwy. 50 on the east and the Otoe/Johnson County boundary (Otoe Rd. S/Johnson Rd. 738 intersection) on the north.

## Other Remnant or Restored Prairies in Eastern Nebraska

Anyone wanting to observe tallgrass prairie birds and native plants should consider visiting some of the following tallgrass prairies. Most are mapped and described in greater detail in *A Nebraska Bird-finding Guide* (Johnsgard, 2005). Many Nebraska prairies are also described in the *Prairie Directory of North America* (Adelman and Schwartz, 2013). Wildlife management areas (WMAs) are Nebraska Game and Parks Commision (NGPC) sites and have free access, but state recreation areas (SRAs) and state parks have daily or seasonal entrance fees. Latitude and longitude (Lat/long) coordinates are shown for most of these state-owned sites. They and other public-access state and federal sites are mapped in the *Public Access Atlas*, published annually by the Nebraska Game and Parks Commission (http://outdoornebraska. gov/publicaccessatlas/). Some sanctuaries and national wildlife refuges charge entry fees. Conservation easement sites require permission from the owner to visit, as do some of the Nature Conservancy (TNC), Prairie Plains Resource Institute (PPRI) and Platte River Recovery Implementation Program (PRRIP) sites.

### Antelope County

#### Grove Lake WMA

2009 acres. Mostly mixed-grass upland and Sandhills prairie and riparian hardwoods along East Verdigre Creek. A stand of tallgrass prairie on sand and gravel is located 100 yards northeast of the parking area. NGPC: 402-471-0641. Lat 42.36, long −98.11.

### Boone County

#### Olson Nature Preserve

112 acres. Sandhills prairie and oak woodlands. Located 8 miles north of Albion on Hwy. 14, then west one mile.

For more information contact PPRI at 402-694-5535 or see http://www.prairieplains.org/preserves/. Lat 41.80, long −98.11.

### Buffalo County

#### Lillian Annette Rowe Bird Sanctuary and Iain Nicolson Audubon Center

2,400 acres. This Audubon sanctuary has six miles of Platte River frontage, with 420 acres of native prairie and 220 acres of restored prairie. It is located 2 miles south of Gibbon at I-80 exit 285, then two miles west on Elm Island Rd. (about 500 yards south of the Platte River bridge). National Audubon Society: 308-468-5282.

#### Pearl Harbor Survivors Preserve

320 acres. Virgin prairie and cropland. North of Riverdale on Pole Line Rd. Used by the University of Nebraska–Kearney as a field laboratory. For more information contact PPRI at 402-694-5535 or see http://www.prairieplains.org/preserves/. Lat 40.88, long −99.19.

### Butler County

#### Timber Point Lake WMA

160 acres. A reservoir, woodlands, and a substantial prairie area. One mile east, one mile south, and one mile east of Brainard off Co. Rd. V. Lower Platte Natural Resources District: 402-476-2729. Lat 42.16, long −96.97.

### Cass County

#### Platte River State Park

519 acres. Riverine hardwood forest with some tallgrass prairie. West of Louisville at 346th St. NGPC: 402-234-2217.

### Cedar County

#### Wiseman WMA

370 acres. Virgin upland prairie on ridges and hilltops. Located just south of the Missouri River, this area includes

steep wooded loess bluffs with bur oak and grassy ridges. One mile north and 5 miles east of Wynot. NGPC: 402-471-0641. Lat 42.76, long -97.10.

## Colfax County

### Frank L. and Lillian Pokorny Memorial Prairie

40 acres. About 20 acres of virgin tall-grass prairie and a 20-acre prairie resto-ration. Located 2.5 miles west of Hwy. 15 at P Rd. (about 11 miles north of Schuy-ler). For more information contact PPRI at 402-694-5535 or see http://www.prairieplains.org/preserves/. Lat 41.61, long -97.11.

### Whitetail WMA

216 acres. Cottonwood savanna along Platte River bottomland forest. About 2.5 miles south of Schuyler off Co. Rd. 10. NGPC: 402-471-0641. Lat 41.44, long -97.08.

## Dixon County

### Buckskin Hills WMA

450 acres, approximately. Located 2 miles west and 2 miles south of New-castle. Some virgin prairie is present among 340 acres of grassland, woods, and a 75-acre lake. NGPC: 402-370-3374. Lat 42.63, long -96.92.

### Ponca State Park (and Elk Point Bend WMA)

892 acres (and 627 WMA acres). The park contains small stands of virgin prairie on ridges and hilltops. Mostly forested with mature stands of bur oak, walnut, hack-berry, and elms. Located 3 miles north of Ponca. NGPC: 402-755-2284.

## Douglas County

### Allwine Prairie Preserve

160 acres. Located 12 miles northwest of Omaha. From I-680 in west Omaha drive west on W. Dodge Rd. to 144th St. Turn north and go to State St., then west 0.5

mile to the preserve entrance. For per-mission to visit, call the Department of Biology at the University of Nebraska-Omaha at 402-554-2641.

### Boyer Chute National Wildlife Refuge

3,500 acres. Includes about 2,000 acres of restored prairie and riparian woods. Three miles east of Fort Calhoun on Co. Rd. 34, along the Missouri River. US Fish & Wildlife Service: 402-468-4313.

### Bauermeister Prairie

40 acres. Owned by the City of Omaha and part of Zorinsky Lake Park (738 acres). The east park entrance is on 156th St. midway between Q St. and W. Center Road. Of the two marked entrances from 156th St., go to the south entrance and follow the park road to some parking ar-eas near its end. The prairie lies to the south and west, beyond an arm of Zorin-sky Lake, via a walking trail.

### Davis Prairie

25 acres. Restored prairie, a University of Nebraska–Omaha field study area near Elkhorn. For information call UNO De-partment of Biology at 402-554-2641.

### Audubon Society of Omaha Prairie Preserve

13 acres. Formerly known as Jensen Prai-rie. About half of the acreage is virgin prairie, the other half restored. Located at 6720 Bennington Rd., near the inter-section of 72nd St. and Hwy. 36 (or 72nd and McKinley). Obtain permission to visit from the Audubon Society of Omaha: 402-445-4138. http://audubon-omaha.org/conservation/aso-prairie.html

### Neale Woods Nature Center

30 acres, approximately, of restored loess prairie and 520 acres of hardwoods. 14323 Edith Marie Ave., Omaha. Fon-tenelle Forest: 402-731-3140. There is an admission fee.

### Stolley Prairie

24 acres. Owned by the City of Omaha (Northwest Park). Located along the east side of 168th St., midway between Blondo and Dodge St.

## Gage County

### Diamond Lake WMA

360 acres. A 33-acre reservoir with hardwoods and some prairie present. Located 4 miles west of Odell on Hwy. 8. NGPC: 402-471-0641. Lat 40.05, long −96.87.

### Homestead Prairie

195 acres. Located 4.5 miles west of Beatrice on Hwy. 4 at Homestead National Monument of America. About 100 acres of restored prairie on a historic homestead site. Includes a 2.5-mile trail through riparian wooded habitats and restored prairie. A local plant list is available. National Park Service: 402-223-3514.

## Hall County

### The Crane Trust

3,000 acres, approximately. With more than 2,000 acres of native and re-seeded wet meadows, the largest wet meadow acreage in the Platte River valley. About 1.5 miles south of the I-80 Alda exit, on Whooping Crane Dr. via Sandhill Crane Dr. Permission to visit is required. Call the Trust office at 308-384 4633. The Trust's Nature and Visitor Center at I-80 exit 305 has a hiking trail through about 250 acres of tallgrass prairie (seasonally accessible). Contact the visitor center at 308-382-1820. https://cranetrust.org/

## Hamilton County

### Deep Well WMA

238 acres. Includes 43 acres of prairie and prairie restorations, and 60 wetland acres. Three miles south of Phillips on S "D" Rd. NGPC: 402-471-0641. Lat 40.85, long −98.22.

### Gjerloff Prairie

390 acres. Loess prairie and Platte River frontage. Drive 4 miles west of Hwy. 14 from the Marquette corner, then 0.5 mile north. Owned by PPRI: 402-694-5535. http://www.prairie-plains.org/preserves/. Lat 41.01, long −98.07.

### Lincoln Creek Prairie and Trail

16 acres. Tallgrass prairie and prairie restorations. Located in Aurora. PPRI: 402-694-5535. http://www.prairie-plains.org/preserves/. Lat 40.87, long −97.99.

### Marie Ratzlaff Prairie Preserve

40 acres. Upland prairie (30 virgin acres). Located 6 miles south of the Hampton I-80 exit (exit 332), west side of the road. PPRI: 402-694-6635. http://www.prairieplains.org/preserves/. Permission to visit not necessary. Lat 40.74, long −97.88.

### Nelson Waterfowl Production Area

162 acres. Includes 145 acres of wetland and 17 acres of upland prairie. Located 2.5 miles north of Stockham on E 5 Rd. US Fish & Wildlife Service: 308-263-3000. Lat 40.76, long −97.94.

### Pintail WMA

478 acres. Includes 268 acres of wetland and 25 acres of prairie pasture. Five miles south and 2 miles east of Aurora at the intersection of E 7 Rd. and S "S" Rd., or 5 miles north of Stockham on E 5 Rd. NGPC: 402-471-0641. Lat 40.79, long −97.95.

### Springer Waterfowl Production Area

640 acres. Wetlands totaling 397 acres plus 160 acres of grassland restoration. Five miles north of Giltner and 1 mile

east on West 11 Rd. US Fish & Wildlife Service: 308-263-3000. Lat 40.85, long -98.13.

**Troester Basin Waterfowl Production Area**

317 acres. Includes 271 acres of wetlands and 49 acres of upland grassland. Four miles south of Aurora and 3.5 miles east on E 8 Rd. US Fish & Wildlife Service: 308-263-3000. Lat 40.97, long -97.92.

## Jefferson County

**Rock Creek Station State Historical Park**

350 acres. Virgin tallgrass prairie on hilltops with wooded ravines. Six miles east of Fairbury and 2 miles south on 574 Ave. A park entry permit is required for the park and adjoining SRA. NGPC: 402-729-5777. Lat 40.11, long -97.06.

**Rock Glen WMA**

706 acres. Includes nearly 500 acres of virgin upland and restored prairie. Located 7 miles east and 2 miles south of Fairbury, or 4 miles northeast of Endicott. NGPC: 402-749-7650. Lat 40.10, long -97.06.

**Rose Creek WMA**

384 acres. About 200 acres of oak savanna. Eight miles southwest of Fairbury on 708 Rd. NGPC: 402-749-7650. Lat 40.08, long -97.24.

## Johnson County

**Hickory Ridge WMA**

250 acres. Includes 17 acres of prairie restoration. Ten miles south of Sterling on 611 Ave. NGPC: 402-471-0641. Lat 40.31, long -96.36.

**Twin Oaks WMA**

1,118 acres. Mostly lowland woods, but the acreage includes some native and restored prairie. One mile south and

3 miles east of Tecumseh on 623 Ave. NGPC: 402-471-0641. Lat 40.33, long -96.14.

## Kearney County

**Spiedell Island Preserve**

596 acres. A TNC prairie preserve on Dover Island, Platte River. Southeast of Kearney. Contact the Nature Conservancy for information about access: 402-343-0282.

## Knox County

**Bohemia Prairie WMA**

680 acres. Nearly 600 acres of virgin prairie, with some woods and two ponds. From Verdigre go 5 miles west on Hwy. 84 and then 5 miles north. NGPC: 402-370-3374. Lat 42.68, long -98.13.

**Greenvale WMA**

200 acres. About 70 acres of virgin prairie almong Middle Verdigre Creek woodlands. Located 10 miles west and 3 miles south of Verdigre. NGPC: 402-370-3374. Lat 42.54, long -98.22.

**Niobrara State Park**

1,632 acres. Grasslands and riparian woods. Located at the west edge of Niobrara. A state park entry permit is required. Inquire at the park office for the prairie site locations. NGPC: 402-857-3373.

## Lancaster County

**Branched Oak SRA**

1,031 acres. About 200 acres of virgin prairie, plus restored prairie. Some good prairie is located below the dam near the main south entrance. Located about 4 miles west of Raymond on Raymond Rd. State park entry permit is required. NGPC: 402-783-3404. Lat 40.98, long -96.88.

**Capitol Beach Saline Wetlands**
30 acres, approximately. The site of a historic saline lake, the east edge of Capitol Beach still supports a saline marsh habitat and associated low prairie. Enter Westgate Blvd. from Sun Valley Blvd., then proceed west on W Industrial Lake Dr. to the parking area. Owned by the Lower Platte South NRD: 402-476-2729. https://www.lpsnrd.org/ (Recreation > Wetlands).

**Frank Shoemaker Marsh**
160 acres. Restored prairie wetlands along Little Salt Creek. Two miles north of Lincoln at Bluff Rd. and North 27th St. Lincoln Parks & Recreation: 402-441-7847. http://www.lincoln.ne.gov/city/parks/parksfacilities/wetlands/shoemaker.htm

**Little Salt Creek WMA and Little Salt Creek West WMA**
256 acres and 220 acres, respectively. Saline wetlands and salt-tolerant grasslands. Little Salt Creek WMA is south of Raymond Rd. between N 14th St. and N 1st St. Little Salt Creek West WMA is south of Branched Oak Rd. between N. 1st St. and N 14th St. For information, contact NGPC: 402-471-0641.

**Little Salt Fork Marsh Preserve**
280 acres. Salt marsh and salt-tolerant grasses. From W 1st St. in northwest Lincoln (near exit 401 of I-80), go north 6 miles to Raymond Rd. The preserve is at the northwest corner of Raymond Rd. and W. 1st St. For information, contact the owner, The Nature Conservancy, at 402-342-0362 or the Lower Platte South NRD at 402-476-2729. https://www.nrdnet.org/rec-area/little-salt-fork-marsh-preserve

**Little Salt Springs Recreation Area and Little Salt Fork Marsh Preserve**
123 acres and 240 acres, respectively.

Salt marsh and salt-tolerant grasses. Both of these areas are south of Branched Oak Rd. Little Salt Springs is west of NW 12th St., and Little Salt Fork Marsh is west of N 1st St. For information, contact the Nature Conservancy at 402-342-0282 or the Lower Platte South NRD at 402-476-2729.

**Pawnee Lake State Recreation Area**
2.544 acres. Several acres of native prairie are on the east side of the lake, just south of Superior St., 2 miles north and 1.5 miles west of Emerald. A state park entry permit is required. NGPC: 402-783-3404.

**Pioneers Park**
626 acres. A nature trail extends southwest from the Chet Ager Nature Center into restored and native prairie. The Prairie Center also has restored prairie and a native plant garden. A third area of fine prairie is located east of the golf course via a footpath west of the parking area near the elk statue. Located southwest of West Van Dorn St. and Coddington Ave. Lincoln Parks & Recreation: 402-441-7895.

**Wilderness Park**
1,472 acres. A 7-mile stretch of riparian woodland along Salt Creek on the southwest side of Lincoln. Stands of mature bur oak and hickory, especially at the south end, and riparian forest, with about 20 miles of trails. Includes a small area of prairie on sandstone outcrops at the west edge of the park, 0.2 mile south of Pioneers Blvd. and on the east side of S 1st St. Lincoln Parks & Recreation: 402-441-7895. http://lincoln.ne.gov/city/parks/parksfacilities/parks/wilderness.htm

**Wildwood WMA**
493 acres. 120 acres of virgin prairie and a 107-acre lake. About 2.5 miles south of

Valparaiso, west of Hwy. 79. Lower Platte South NRD: 402-476-2729. Lat 40.09, long –96.84.

## Zoetis Saline Wetlands

25 acres. A footpath travels through restored wetlands and wet prairie. Located on W Cornhusker Hwy. at N 1st St. (Formerly the Pfizer Saline Wetlands.) Free admission. http://www.lincoln.ne.gov/city/parks/parksfacilities/wetlands/zoetis.htm

## Madison County

### Oak Valley WMA

640 acres. Mixed virgin prairie and bur oak draws. Hardwood forest along Battle Creek and prairie uplands. Located 2.5 miles south, 0.5 mile west of Battle Creek. NGPC: 402-675-4020. Lat 41.95, long –97.62.

## Merrick County

### Bader Memorial Park

Overall acreage unavailable. Consists of 0.75 mile of Wood River and 0.5 mile of Platte River frontage. Includes about 120 acres of partly restored grassland as well as riparian forest and shrubland. Located at the west end of the Chapman (Platte River) Bridge. Admission fee. Owned by the county: 308-986-2522. Lat 40.98, long –98.15.

## Nance County

### Olson Nature Preserve

112 acres. Sandhills and tallgrass prairie and oak woodlands. Located 8 miles north of Albion on Hwy. 14, then one mile west on Y Rd. to gate. PPRI: 402-694-5535. Lat 41.80, long –98.11. http://www.prairieplains.org/preserves/

### Sunny Hollow WMA

160 acres. Virgin wet prairie plus two marshes and a wetland. Located 5 miles south of Genoa at 530 St. NGPC: 402-471-0641. Lat 41.38, long –97.74.

## Pawnee County

### Burchard Lake WMA

560 acres. About 400 acres of rolling prairie over limestone. Drive 3 miles east of Burchard, then 1 mile north, or go east 3 miles on Hwy. 4 from the junction of Hwys. 99 and 4 (junction is 3 miles north of Burchard), and then 1.5 miles south. NGPC: 402-471-0641. Lat 40.17, long –96.30.

### Pawnee Prairie WMA

1,120 acres. This WMA is partly prairie. Located 8 miles south of Burchard and 1 mile east via Hwys. 8 and 99. Also accessible by driving 10 miles south of Burchard and 1 mile east. The prairie is to the east of these access points. NGPC: 402-471-0641. Lat 40.03, long –96.37.

### Table Rock WMA

415 acres. Includes some prairie. One mile east of Table Rock on Hwy 4. NGPC: 402-471-0641. Lat 40.18, long –96.06.

## Platte County

### Wilkinson WMA

940 acres. About 80 acres of wet virgin prairie, plus upland grassland and wetlands. Five miles west and 1 mile north of Columbus off Hwy. 81. NGPC: 402-370-3374. Lat 41.50, long –97.49.

## Richardson County

### Indian Cave State Park

3,400 acres. About 40 acres of virgin prairie on hilltops and hay meadows in southeast part of park, and also along Trail 10 from top of bluffs to the Adirondack shelter. Diverse wooded habitats and flora of southern botanic affinities. The park is about 80 percent mature hardwood forest with the rest grassland. Park entry fee. NGPC: 402-883-2575. Lat 40.26, long –95.57.

**Rulo Bluffs Preserve**

445 acres. Ridgetop tallgrass prairie savanna and hardwood forest on high loess bluffs. Located about 1.5 miles south and 4 miles east of Rulo along the west shore of the Missouri River. Very steep terrain. For permission to visit, contact the Nature Conservancy: 402-342-0282.

## Saline County

**Swan Creek WMA**

160 acres. Includes 54 acres of prairie pasture. Seven miles east of Milligan on Hwy. 41. NGPC: 402-471-0641. Lat 40.51, long –97.25.

## Sarpy County

**Fontenelle Forest**

2,000 acres. Mostly mature riparian forest but some oak savanna and prairie. Nature center. At 1111 N. Bellevue Blvd. N, Bellevue, Nebraska. Entrance fee. 402-731-3140.

**Schramm Park SRA**

337 acres. Includes some native prairie. Seven miles south of Gretna on Hwy. 31. Park entrance fee required. NGPC: 402-234-6855.

## Saunders County

**Larkspur WMA**

160 acres. Native prairie and restored grasslands. Four miles west of Valparaiso, 1 mile north of Hwy 66. NGPC: 402-471-0641. Lat 41.09, long –96.90. (On the way, the Valparaiso Cemetery just west of town on Hwy. 66 has a small area of prairie.)

**Madigan Prairie and Red Cedar Lake Recreation Area**

23 acres and 120 acres, respectively. Native and restored grasslands. Both areas are located between Valparaiso and Weston. Red Cedar Lake is 6 miles north and 2 miles west of Valparaiso. Madigan

Prairie is 1 mile east of the Butler County line and 2 miles south of Hwy. 92. The Lower Platte South NRD manages Red Cedar: 402-476-2729. The University of Nebraska Foundation owns Madigan Prairie: 402-472-2083 and http://snr. unl.edu/aboutus/where/fieldsites/madiganprairie.aspx

**Memphis SRA**

163 acres. This recreation area has some restored prairie along with a 48-acre lake. Located 1 mile north and 1 mile west of Memphis off Rd. D. A park entrance permit is required. NGPC: 402-791-5497.

## Seward County

**Bur Oak WMA**

139 acres. Mostly mature bur oak woodland with about 40 acres of prairie in oak savanna. Located 5 miles east of Seward along Hwy. 34. NGPC: 402-471-0641. Lat 40.90, long –97.00.

**Twin Lakes WMA**

1,300 acres. About 600 acres of upland prairie plus two lakes, marshes, and wooded bottomlands. Located 0.5 mile north and 0.5 mile west of I-80 (Pleasant Dale) exit 388. The best prairie is southwest of the smaller lake, on the west side of the WMA. NGPC: 402-471-0641. Lat 40.83, long –96.95.

## Stanton County

**Red Fox WMA**

363 acres. Includes 163 acres of native grassland. Located south of Pilger on Hwy. 15. NGPC: 402-471-0641. Lat 41.99, long –97.05.

**Wood Duck WMA**

668 acres. Virgin prairie on sand-gravel soils and restored prairie. Also riparian wooded habitats and oxbow lakes bordering the Elkhorn River. Located about

2 miles south and 4 miles west of Stanton. NGPC: 402-370-3374. Lat 41.93, long –97.30.

## Thayer County

**Father Hupp WMA**
160 acres. Includes some upland prairie. Two miles west of Bruning on Co. Rd. Z. NGPC: 402-471-0641. Lat 40.34, long –97.62.

**Meridian WMA**
400 acres. Includes 180 acres of upland prairie. Located 3.5 miles north of Gilead, or 1 mile west and 3 miles south of Alexandria. NGPC: 402-749-7650. Lat 40.20, long –97.40.

## Washington County

**Boyer Chute National Wildlife Refuge**
3,500 acres. Riparian woods and 52 acres of restored prairie. Three miles east of Fort Calhoun on Co. Rd. 34. US Fish & Wildlife Service: 402-468-4313.

**Cuming City Cemetery and Nature Preserve**
11 acres. Virgin prairie. From the intersection of US Hwys. 30 and 75 in Blair, go north 3.5 miles on Hwy. 75 to Co. Rd. 14. Turn left and go 600 feet to the cemetery entrance on the left. Once owned by Dana College. Lat 41.59, long –96.17.

**DeSoto National Wildlife Refuge**
8,362 acres (13.07 square miles). About 1,900 acres of restored prairie among riparian woods along with an oxbow lake and wetlands. Five miles east of Blair on Hwy. 30. Entrance fee. US Fish & Wildlife Service: 712-642-4121.

**Fort Atkinson State Historical Park**
157 acres. Includes some restored prairie. Located at the eastern edge of Fort Calhoun. NGPC: 402-468-5611.

## Wayne County

**Thompson-Barnes WMA**
160 acres. About 18 acres of restored prairie. From Wayne go 3.5 miles north on Hwy. 15 and 1 mile west. NGPC: 402-370-3374. Lat 42.30, long –97.04.

*My own feeling for tallgrass prairie is that of a modern man fallen in love with the face in a faded tintype. Only the frame is still real; the rest is illusion and dream. So it is witih the original prairie. The beautiful face of it had faded before I was born, before I had a chance to touch and feel it, and all that I have known of the prairie is the setting and the mood—*

John Madson, *Where the Sky Began: Land of the Tallgrass Prairie*

And yet these prairie remnants can tell us hints of what was, and even if forever gone, can instill in us authentic new memories of the beauties of our earth, of our lives, and of the lives of those who in the future will continue to treasure, protect, and nourish these tiny treasures that silently speak to us of our collective past.

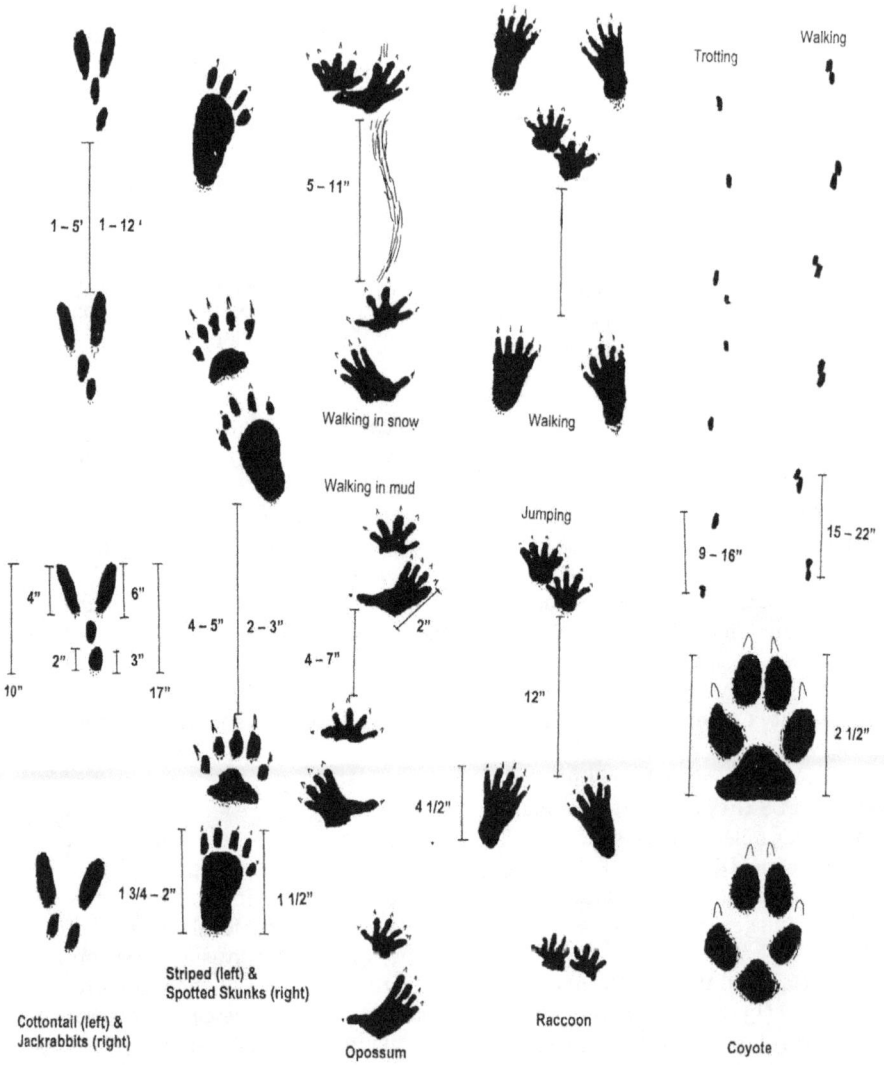

Trotting

Walking

5 – 11"

1 – 5'  1 – 12'

Walking in snow

Walking

Walking in mud

Jumping

9 – 16"

15 – 22"

4"          6"

4 – 5"  2 – 3"

2"      3"

4 – 7"

12"

2 1/2"

10"

17"

2"

4 1/2"

1 3/4 – 2"     1 1/2"

Striped (left) &
Spotted Skunks (right)

Cottontail (left) &
Jackrabbits (right)

Opossum

Raccoon

Coyote

*Fig. 41. Footprints of mammals of eastern Nebraska*

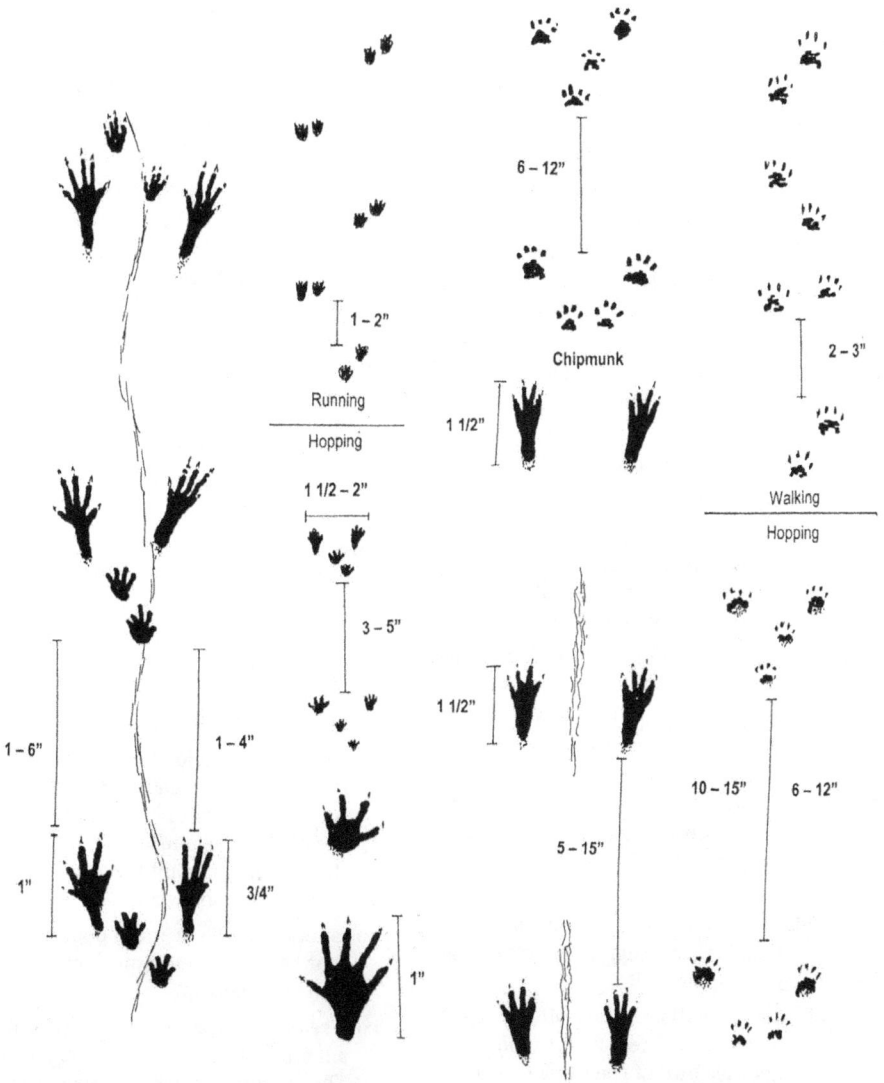

6 – 12"

Chipmunk

1 1/2"

1 – 2"

Running

Hopping

1 1/2 – 2"

3 – 5"

1 1/2"

1 – 6"

1 – 4"

1"

3/4"

2 – 3"

Walking

Hopping

10 – 15"

6 – 12"

5 – 15"

1"

Jumping Mouse (left)
& Pocket Mouse (right)

White-footed Mouse
& Deer Mouse

Kangaroo Rat

Franklin's (left) & 13-lined
(right) Ground Squirrels

# Glossary

**accipiter**  A member of the hawk genus *Accipiter*.

**adaptation**  A genetic trait that increases the ability of an individual organism to better survive and reproduce within its environment, and is thus favored (selected for) by natural selection.

**allopatric**  Descriptive of two populations having nonoverlapping geographic ranges. See also sympatric.

**andromorph**  Taking the form of a male. *See also* heteromorph.

**annual**  A plant or other organism that matures, reproduces, and dies within a single year or growing season.

**biennial**  A plant or other organism that requires two years to mature, reproduce, and complete its life cycle.

**biomass**  The total living weight (mass) of a species or a collection of species that occupy a particular location.

**biome**  A geographically large-scale ecosystem of plants, animals, and their environment that share important common characteristics (e.g., climate, soil, dominant species), such as the tallgrass prairie biome.

**biota**  All the living organisms of a particular place, region, or some larger geographic entity.

**brood parasitism**  The laying of eggs by an individual female into the nest of another bird of the same or a different species. Also called egg parasitism and nest parasitism.

**buteo**  A member of the hawk genus *Buteo*.

**carapace**  The dorsal bony "shell" of a turtle.

**Cenozoic era**  The geologic period encompassing the past 65 million years, the so-called Age of Mammals.

**chrysalis**  (chrysalides, chrysalises, pl.) The hardened outer protective layer of a pupating butterfly. *See also* cocoon, larva, pupa.

**class**  In taxonomy, the category above the order and below the phylum. *See also* family, order.

**clone**  Offspring that are genetically identical to their parents, such as those developing from unfertilized eggs, a phenomenon called parthenogenesis. Cloning is the process of producing clones.

**cocoon**  An insect pupa that is encased in a larval-produced silken envelope. *See also* chrysalis, larva, pupa.

**coevolution**  The reciprocal evolutionary influences of two interactive species on one another over time, typically to their mutual (symbiotic) benefit.

**community**  In ecology, interacting populations of plants and animals that occupy a specific site. Plant communities are often named after one or two of their most dominant species. *See also* ecosystem, dominant.

**conspecific**  Two or more populations that belong to the same species. *See also* interspecific.

**cool-season grasses**  Grasses genetically adapted for growing in cool climates. They require more water than warm-season grasses. Also called C-3 grasses because of an intermediate 3-carbon-molecule stage present during photosynthesis. *See also* warm-season grasses.

**cranial crest**  Bony enlargements centered along the dorsal midline of the skull of *Bufo* toads.

154

**crepitation** Crackling noises made by some grasshoppers during flight. *See also* stridulation.

**crepuscular** Active during the periods of dusk and dawn. *See also* diurnal, nocturnal.

**dimorphism** Occurring in two forms or appearances, such as sexual dimorphism. *See also* monomorphism.

**diurnal** Activity occurs during daylight hours. *See also* crepuscular, nocturnal.

**dominant** In plants, those species that have the strongest ecological effects in a biological community and if removed would have the greatest disruption on the remaining community.

**double brooding** The completion of two nesting cycles during a single season. Sometimes called multiple brooding. *See also* renesting.

**ecosystem** An interacting group of plants, animals, and their physical and chemical environment, as limited by energy flows and nutrient cycles. Ecosystems might be as small and transient as a mud puddle or as large, complex, and permanent as the earth. *See also* community.

**ecotone** An ecological transition zone, such as a prairie-woodland transition.

**endangered** Descriptive of taxa that exist in such small numbers as to be in direct danger of extinction.

**endemic** Refers to a population native to and confined to a particular area or region. *See also* pandemic.

**exotics** Taxa that have been accidentally or purposefully introduced into an area where they are not native.

**extirpated** Descriptive of taxa (usually a species or subspecies) that has been locally eliminated from an area or region but still persists elsewhere within its overall range.

**facultative** Refers to unconstrained activities or behavior, such as being able to parasitize or exploit a variety of other organisms, rather than a single kind. *See also* obligatory.

**family** In taxonomy, the category above that of the genus (or the subfamily, if present) and below that of the order. In animal nomenclature, family names consistently end in "idae" and subfamily names end in "inae." *See also* genus, order.

**fauna** The collective animal life of a defined area or region.

**fledging period** The amount of time required for an individual bird to pass from hatching to its first flight. *See also* nestling period.

**flora** The collective plant life of a defined area or region. A flora might also be a listing or description of all the species that compose a particular taxonomic group, or those that occur in a specific region.

**genotype** The genetic makeup of an individual plant or animal. *See also* phenotype.

**genus** (genera, pl.; generic, adj.) One or more species that have certain shared characteristics that indicate they are a closely related evolutionary assemblage. The genus is also the nomenclatural taxonomic category immediately above the species and below the family levels, and when printed is italicized and capitalized. *See also* family, species.

**glycoside** Any of various compounds found in plants that yield a sugar on hydrolysis.

**guild** A group of species that occupy the same habitat and share certain important niche characteristics. *See also* niche.

**habitat**  The physical and biological environment of a specific place or general area.

**herpetile**  ("herps," informally) A collective term for reptiles and amphibians. Herpetology is the study of herpetiles.

**heteromorph**  Taking the form of the opposite sex. *See also* andromorph.

**hibernaculum**  An overwintering site.

**home range**  The area occupied but not necessarily defended by a mobile animal over a stipulated time period. *See also* territoriality.

**hybrid**  The offspring of a mating between two species. Offspring of matings between subspecies are often called genetic intergrades rather than hybrids. *See also* species, subspecies.

**infrared**  Those light waves at the red end of the visible spectrum that are too long to be perceived as light ("below red") but might be detected as heat. *See also* ultraviolet.

**interspecific**  Refers to interactions between two or more species, such as interspecific hybridization. *See also* conspecific.

**larva**  (larvae, pl.) A free-living embryonic stage in the life cycle of an animal that is structurally different from that of the adult stage, which is attained by a gradual process of transformation (metamorphosis).

**larynx**  The vocal structure of mammals, located between the pharynx and trachea, and containing the vocal cords. *See also* pharynx, syrinx.

**lek**  A location at which sexually active males of a local mating population meet to compete with one another for access to mating opportunities, either by establishing social dominance over other males, or being able to effectively attract females. Male behavior at leks is called lekking; leks are sometimes also called arenas.

**leucistic**  (leucism, n.) A variant (morph) plumage in birds in which the pigment pattern is very pale because unusually little melanin is present. *See also* melanistic. Plumages that have only one of two normally present pigments (such as melanins and carotinoids) are sometimes described as schizochroistic.

**lore**  (lores, pl.; loral, adj.) In ornithology, refers to the part of the head that is between the eye and the beak or bill.

**malar**  In ornithology, refers to the part of the head that would be the lower cheeks of a mammal, or above the throat and below the lores in birds. Linear feather markings here are often called malar stripes or "mustaches." *See also* lore.

**melanistic**  (melanism, n.) A variant (morph) plumage in birds in which the plumage pattern is very dark because unusually large amounts of melanin are present. *See also* leucistic.

**mesic**  Environmental moisture conditions intermediate between dry (xeric) and wet (hydric). *See also* xeric.

**metamorphosis**  A process of structural or anatomical change in an animal's life cycle, typically from a larva to an adult, which in some animals is marked by an intervening quiescent ("pupal") period. Some insects (such as the butterflies) during pupation undergo complete metamorphosis, which is described as holometamorphosis; other insect groups (such as the grasshoppers) exhibit continuous graduated stages of development to an adult condition through a progressive series of molts, a process called incomplete metamorphosis, or hemimetamorphosis. *See also* larva, pupa.

**mixed-grass prairie**  (or midgrass prairie) Perennial grasslands that are dominated by intermediate-stature grasses, often 0.5–1.5 meters high. These typically occur in enviroments that are more mesic (relatively moist) than those supporting only short grasses, but too xeric (moisture-deficient) to support tall grasses. *See also* short-grass prairie, tallgrass prairie.

**monogamous**  A mating system in which females and males pair-bond and mate with a single member of the opposite sex, either over a single breeding cycle or indefinitely longer. *See also* polyandrous, polygynous, promiscuous.

**monomorphism**  Occurring as a single form or appearance, such as sexual monomorphism, in which the sexes are externally identical (monomorphic). *See also* dimorphism.

**morph**  A specific gene-based phenotype of a population in which two or more different and stable phenotypes (morphs) occur. Synonymous with the more commonly used term "phase," which inaccurately implies a temporary or seasonal phenotypic condition. *See also* genotype, phenotype.

**mutualism**  An ecological situation in which two species mutually benefit from each other's presence, which might involve facultative or obligatory interactions. Often also called symbiosis. *See also* coevolution.

**neotony**  (neotenous, adj.) The retention (in some salamanders) of larval structures such as external gills and fins into sexual adulthood.

**nesting period**  As used here, the duration of a species' or individual's breeding time from the start of the first nest until the last nest is abandoned.

**nestling period**  The amount of time an individual young bird is a nestling, from hatching until it leaves the nest.

**niche**  An organism's specific role or biological "profession" within its ecological environment, as defined by its behavioral, morphological, and physiological adaptations to that environment. *See also* adaptation, habitat.

**nocturnal**  Activity occurs during nighttime. *See also* crepuscular, diurnal.

**nomenclature**  The process of naming things. Biological nomenclature consists of the naming and describing of organisms. *See also* systematics.

**nonpasserine**  A collective term for all the birds that do not belong to the order Passeriformes (the so-called songbirds). *See also* passerine.

**obligatory**  Refers to constrained activities or behavior, such as an obligatory host species or symbiotic partner that requires the presence of another species. *See also* facultative.

**order**  In taxonomy, the category above that of the family and below that of the class. In animal nomenclature, at least in bird taxonomy, the names of orders consistently end in "iformes," as in Passeriformes; in other vertebrate groups ordinal names often end in "ia," as in Rodentia and Lacertilia. *See also* class, family.

**oviparous**  Descriptive of animals that give birth to shelled embryos (eggs) that complete their development outside the mother's body. When the development of the embryos that occurs within the eggs occurs inside the mother's body until they are ready to hatch, the reproductive process is called ovovivipary. *See also* viviparous.

**ovoposition**  The act of laying (ovopositing) a shelled egg externally. The internal release of an unfertilized ovum is ovulation.

**pandemic**  Refers to a population that occurs widely over a region, continent, or the world. *See also* endemic.

**parotoid gland**  A large secretory gland located behind each eye in *Bufo* toads.

**parthenogenesis**  Reproduction of bisexual animals by females without prior male fertilization (so-called "virgin birth"). *See also* clone.

**passerine** A collective term for all the birds that belong to the order Passeriformes, the so-called songbirds. *See also* nonpasserine.

**perennial** A plant of indefinite lifespan.

**pharynx** The part of the digestive tract that extends from the mouth to the larynx, where it joins the esophagus. *See also* larynx.

**phenotype** The external appearance of an individual plant or animal. *See also* genotype.

**pheromones** Chemicals that are produced by animal, released into the air, and received by other conspecific individuals, on which they have physiological or behavioral effects. In some plants somewhat similar chemicals (allomones) are produced by roots or fallen leaves that influence the growth or survival of other plants that are nearby.

**plastron** The ventral bony "shell" of a turtle.

**Pleistocene epoch** The geologic time interval ("epoch") extending from about 1.6 million years ago (the end of the Pliocene epoch) to 12,000 years ago (the start of the Holocene epoch). Popularly known as the "Ice age," this epoch included several major glacial periods and associated interglacial intervals, of which the Kansan glaciation (ca. 1.4–0.9 million years ago) was the only glacier that extended to southeastern Nebraska and northeastern Kansas.

**pollinaria** Pollen clusters of orchids.

**polyandrous** A mating system in which females may pair-bond and mate with two or more males simultaneously or successively. *See also* monogamous, polygynous, promiscuous.

**polygynous** A mating system in which males may pair-bond and mate with two or more females simultaneously or successively. *See also* monogamous, polyandrous, promiscuous.

**prairie** A native plant community that is dominated by perennial grasses. Prairies may be further defined by the relative stature of the dominant species (tallgrass, mixed-grass, shortgrass), by their dominant plant taxa (e.g., big bluestem), or by the life form of these grasses (bunchgrasses versus sod-forming grasses).

**promiscuous** A mating system in which both sexes may mate with multiple members of the opposite sexes, often without any other prolonged social interactions. *See also* monogamous, polyandrous.

**pupa** (pupae, pl.) The external shell of an insect undergoing metamorphism from the larval to adult stage. In some insects the pupae are covered by silken coats, or cocoons. *See also* chrysalis, cocoon.

**refugium** (refugia, pl.) A location or region where certain taxa have been able to survive after the general environment or climate has changed and eliminated those taxa from surrounding areas. *See also* relicts.

**relicts** Plant or animal taxa persisting locally in an otherwise markedly altered habitat or climate. *See also* refugium.

**renesting** A second breeding effort undertaken following the failure of a first attempt. *See also* double brooding.

**rhizome** (rhizomatous, adj.) A horizontal underground root-like stem. *See also* tiller.

**riparian** Descriptive of a bank habitat beside water, such as a shoreline or riverbank plant community. "Riverine" is a similar term, referring to a river-related habitat specifically.

**ruderal** Refers to plants growing in waste places.

**sedge** A general term for members of the plant family Cyperaceae, which are grass-like plants having solid, triangular stems, rather than the round and hollow stems typical of grasses (family Poaceae).

**shortgrass prairie** Perennial grasslands that occur in regions too dry to support mixed-grass prairies. *See also* mixed-grass prairie, steppe.

**song** In birds, complex vocalizations that are produced by paired vibratory membranes in the syrinx and are variably modulated in frequency, duration, loudness, and harmonic content by the trachea, pharynx, oral cavity, and in some birds also the upper esophagus. *See also* passerine, pharynx, songbird, syrinx.

**songbird** An informal term that refers to all "perching" birds. Collectively, they have long hind toes that permit perching on branches and so forth, and they belong to the order Passeriformes, or informally, "passerine birds." Not all passerines are true songbirds. The more "primitive" passerines, such as kingbirds and other tyrannids (family Tyrannidae), are often highly vocal but lack the complex vocal (syringeal) structures of the more advanced songbirds that produce acoustically complex songs. The latter group makes up the oscines, versus the more primitive passerines, which are called suboscines. *See also* passerine, syrinx.

**species** (specific, adj.) Informally, a population of plants or animals having definable and stable morphological traits; more accurately, a population of individuals capable of interbreeding with others of that population but reproductively isolated from all other populations. The term "species" is also a taxonomic category, which is below that of the genus and above the subspecies category. The Latin names of species and subspecies are italicized but are not capitalized, but generic names (e.g., *Passer*) are always italicized and capitalized. When written, the organism's specific Latin name follows the generic name (e.g., *Passer domesticus*). The categories at higher taxonomic levels, such as families (e.g., Passeridae) and orders (e.g., Passeriformes) are capitalized but not italicized. *See also* family, genus, order, subspecies.

**steppe** A Russian term for shortgrass prairie. *See also* shortgrass prairie.

**stridulation** Mechanical noises made by insects and other arthropods by rubbing parts of their exoskeleton together. *See also* crepitation.

**subspecies** Defined populations of a species that are geographically separated (allopatric) but not genetically isolated from other such populations. Subspecies are often called races. *See also* allopatric.

**succession** In ecology, refers to changes in the botanical and zoological composition of the plant and animal life of a biological community over time, from early ("pioneer") through transitional ("seral") stages until a stabilized ("climax") biotic composition is attained, as influenced by climate, substrate, and other limiting factors.

**symbiosis** The association of two organisms of different species to the benefit of one or both species. *See also* coevolution, mutualism.

**sympatric** Descriptive of two species having significant geographic overlap, at least during their breeding season. Also called syntopic. *See also* allopatric.

**syrinx** (syringeal, adj.) The unique vocal structure of birds, located at the junction of the two bronchi and the trachea, and usually consisting of two pairs of vibratory membranes (tympanic membranes). *See also* larynx.

**systematics** Biological systematics consists of the erection of nomenclatural categories that best reflect the apparent evolutionary relationships of two or more groups of organisms. *See also* taxonomy.

**tallgrass prairie** ("true" prairie) Perennial grasslands that are dominated by tall-stature grasses, often at least two meters high. These typically occur in climates that are more mesic (relatively moist) than those supporting intermediate-height grasses but too xeric (moisture-deficient)—or too frequently burned—to support forests. *See also* mixed-grass prairie.

**taxon** (taxa, pl.) A biological unit in the taxonomic hierarchy, from subspecies to kingdom. Taxonomy is the study and systematic organization of taxa according to their evolutionary relationships (phylogenies), including their technical names (taxonomic nomenclature).

**taxonomy** (taxonomic, adj.) The process of systematically organizing organisms in an orderly series of standardized categories that best reflect their evidence-based evolutionary relationships. Also called systematics. *See also* nomenclature.

**terpene** Any of various unsaturated hydrocarbons, such as essential oils, that are produced by some plants.

**territoriality** The advertisement and defense of an area by an individual of one sex (usually the male) against others of the same sex and species, while simultaneously attracting the opposite sex of that species.

**tiller** An above-ground horizontal stem. *See also* rhizome.

**tympanum** The eardrum of vertebrates, which in amphibians is located on the lateral head surface, behind the eye, and in reptiles (except for snakes), birds, and mammals is located at the end of an auditory canal.

**ultrasonic** Sound frequencies that are above the range of human hearing, generally those above approximately 15,000–20,000 cycles per second (Hz).

**ultraviolet** (UV) Light waves at the violet end of the spectrum that are too short to be seen by humans ("beyond violet"). *See also* infrared.

**vascular plants** A collective term for plants that have more complex internal structures (such as a nutrient transportation system) than do "lower" plants, including mosses and algae.

**viviparous** Descriptive of animals that give birth to living offspring (larvae, etc.) rather than producing eggs that develop outside the body. *See also* oviparous.

**vocalizations** Sounds made by animals through specialized structures for moving air past vibratory membranes (mainly the syrinx in birds, larynx in mammals and amphibians). Vocalizations are usually "formed" sounds, with varied aspects of modulation, sound frequencies, loudness, cadence, and harmonic structure that allow for complex information transfer. *See also* crepitation, larynx, stridulation, syrinx.

**warm-season grasses** Grasses genetically adapted for growing in warm, dry climates and requiring less water than warm-season grasses. Also called C-4 grasses, in reference to an intermediate 4-carbon molecule stage present during photosynthesis. *See also* cool-season grasses.

**xeric** Dry environmental conditions. *See also* mesic.

# References

### General Topics and Plant Ecology

Adelmann, C., and B. L. Schwartz. 2013. *Prairie Directory of North America: The United States, Canada and Mexico.* New York: Oxford University Press. 352 pp.

Allen. D. 1967. *The Life of Prairies and Plains.* New York: McGraw Hill.

Axelrod, D. I. 1985. Rise of the grassland biome, central North America. *Botanical Review* 51: 163–201.

Bailey, R. G. 1995. *Descriptions of the Ecoregions of the United States.* Washington, DC: US Department of Agriculture.

Barkley, T. M., ed. 1977. *Atlas of the Flora of the Great Plains.* Great Plains Flora Association. Ames: Iowa State University Press. 600 pp.

———. 1986. *Flora of the Great Plains.* Great Plains Flora Association. Lawrence: University Press of Kansas. 1392 pp.

Bichel, M. A. 1959. Investigations of a Nebraska and Colorado prairie and their impact on the relict concept. PhD diss., University of Nebraska, Lincoln. 180 pp.

Boettcher, J. F. 1981. Native tallgrass prairie remnants of eastern Nebraska: Floristics and effects of management, topography, size and season of evaluation. MS thesis, University of Nebraska–Omaha, Omaha. 67 pp.

Boettcher, J. F., T. B. Bragg, and D. M. Sutherland. Floristic diversity in ten tallgrass prairie remnants of eastern Nebraska. *Transactions of the Nebraska Academy of Sciences* 20: 33–34. (217 species, including 30 nonnative species)

Brown, L. 1985. *Grasslands.* New York: Knopf.

Carpenter, J. R. 1940. The grassland biome. *Ecological Monographs* 10: 617–684.

Chadwick, D. H. 1993. The American prairie. *National Geographic* 184: 90–119.

Costello. D. F. 1969. *The Prairie World.* New York: T. Y. Crowell.

Cushman, R. C., and S. R. Jones. 2004. *The North American Prairie.* Boston: Houghton Mifflin. 510 pp.

Darwin, C. 1877. *The Various Contrivances by Which Orchids are Pollinated by Insects.* 2nd ed. London: John Murray.

Davies, R. W. 1998. The pollination ecology of *Cypripedium acaule. Rhodora* 88: 445–450.

Duncan, P. D. 1978. *Tallgrass Prairie: The Inland Sea.* Kansas City, MO: Lowell Press. 113 pp.

Farney, D. 1980. The tallgrass prairie: Can it be saved? *National Geographic* 157 (January), pp. 36–61.

Forsberg, M., D. O'Brien, D. Wishart, and T. Kooser. 2009. *Great Plains: America's Lingering Wild.* Chicago: University of Chicago Press. 256 pp.

Gentile, R. J. 2015. *Rocks and Fossils of the Central United States, with Special Emphasis on the Greater Kansas City Area.* 2nd ed. University of Kansas, Department of Geology and Paleontological Institute, Special Publication 8. 221 pp.

Hansen, T., and P. A. Johnsgard. 2007. *Prairie Suite: A Celebration.* Denton, NE: Spring Creek Prairie Audubon Center. 64 pp.

Johnsgard, P. A. 1976. The grassy heartland. Pp. 234–263, in *Our Continent: A Natural History of North America*. Washington, DC: National Geographic Society.

——. 1995. *This Fragile Land: A Natural History of the Nebraska Sandhills*. Lincoln: University of Nebraska Press. 256 pp.

——. 2001a. *Prairie Birds: Fragile Splendor in the Great Plains*. Lawrence: University Press of Kansas. 331 pp.

——. 2001b. *The Nature of Nebraska: Ecology and Biodiversity*. Lincoln: University of Nebraska Press. 402 pp.

——. 2003. *Faces of the Great Plains: Prairie Wildlife*. Lawrence: University Press of Kansas. 190 pp.

——. 2008. *A Guide to the Tallgrass Prairies of Eastern Nebraska and Adjacent States*. University of Nebraska–Lincoln DigitalCommons. 150 pp. http://digitalcommons.unl.edu/biosciornithology/39/

——. 2009a. Forbs and grasses and Cheshire cats: What is a tallgrass prairie? *Prairie Fire*, December, pp. 3, 9. http://www.prairiefirenewspaper.com/2009/12/forbs-and-grasses-and-cheshire-cats-what-is-a-tall-grass-prairie

——. 2009b. Autumn on the prairie: Nebraska's grasses. *Nebraska Life*, September/October, pp. 18–21.

——. 2012a. *A Prairie's Not Scary*. University of Nebraska–Lincoln DigitalCommons and Zea Books. 48 pp. http://digitalcommons.unl.edu/zeabook/10/ (Children's poetry and drawings)

——. 2012b. Spring Creek Prairie Audubon Center: An 800-acre schoolhouse. *Prairie Fire*, October, pp. 18–20, 22. http://www.prairiefirenewspaper.com/2012/10/spring-creek-prairie-audubon-center-an-800-acre-schoolhouse

——. 2012c. *Wetland Birds of the Central Plains: South Dakota, Nebraska, and Kansas*. University of Nebraska–Lincoln DigitalCommons and Zea Books. 275 pp. http://digitalcommons.unl.edu/zeabook/8/

——. 2014. *Seasons of the Tallgrass Prairie: A Nebraska Year*. Lincoln: University of Nebraska Press. 171 pp.

——. 2018a. *A Naturalist's Guide to the Great Plains*. University of Nebraska–Lincoln Digital Commons and Zea Books. 165 pp. http://digitalcommons.unl.edu/zeabook/63/

Johnson, S. R., and A. Bouzaher, eds. 1996. *Conservation of Great Plains Ecosystems*. Boston: Kluwer. 435 pp.

Jones, S. R. 2000. *The Last Prairie: A Sandhills Journey*. New York: Ragged Mountain Press/McGraw Hill.

Kaul, R. B., and S. B. Rolfsmeier. 1987. The characteristics and phytogeographical affinities of the flora of Nine-mile Prairie, a western tallgrass prairie in Nebraska. *Transactions of the Nebraska Academy of Sciences* 15: 23–35.

Knopf, F. L., and F. B. Samson, eds. 1997. *Ecology and Conservation of Great Plains Vertebrates*. Ecological Studies, Vol. 125. New York: Springer.

Kottas, K. L. 2000. Floristic composition and characteristics of Spring Creek Prairie. MS thesis, University of Nebraska–Lincoln.

——. 2001. Comparative floristic diversity of Spring Creek and Nine-mile prairies, Nebraska. *Transactions of the Nebraska Academy of Sciences and Affiliated Societies* 27: 31–59.

Küchler, A. W. 1964. *Potential Natural Vegetation of the Conterminous United States*. Special publication no. 36. New York: American Geographical Society.

Larabee, A., and J. Altman. 2001. *Last Stand of the Tallgrass Prairie*. New

York: Michael Friedman. 144 pp.

Li, P., R. Pemberton, Z. Guiling, and L. Yibo. 2012. Fly pollination in *Cypripedium*: A case study of sympatric *C. sichuanense* and *C. micranthum*. *Botanical Journal of the Linnean Society* 170: 50–58.

Madson, J. 1993. *Tallgrass Prairie*. Helena, MT: Falcon Press. 111 pp.

———. 1995. *Where the Sky Began: Land of the Tallgrass Prairie*. Ames: Iowa State University Press. 340 pp.

Malin, J. C. 1967. *The Grassland of North America: Prolegomena to Its History*. Gloucester, MA: Peter Sith. 398 pp.

———. 1984. *History and Ecology: Studies of the Grasslands*. Lincoln: University of Nebraska Press. 376 pp.

Manning, R. 1955. *Grassland: The History, Biology, Politics, and Promise of the American Prairie*. New York: Viking. 320 pp.

Mantzwe, L. W. 1950. Studies of plant succession in true prairie. PhD diss., University of Nebraska–Lincoln.

Mengel, R. M. 1970. The North American Central Plains as an isolating agent in bird speciation. Pp. 279–340, in *Pleistocene and Recent Environments of the Central Great Plains* (W. Dort, Jr. and J. K. Jones, Jr., eds.). Lawrence: University Press of Kansas.

Moul, F. 2006. *The National Grasslands: A Guide to America's Undiscovered Treasures*. Lincoln: University of Nebraska Press. 155 pp.

Nilsson, L. A. 1992. Orchid pollination biology. *Trends in Ecology and Evolution* 7–8: 255–259.

Omernik, J. M. 1987. Ecoregions of the conterminous United States. *Annals of the Association of American Geographers* 77: 118–125.

Panella, M. J. 2010. *Nebraska's At-risk Wildlife*. Lincoln: Nebraska Game and Parks Commission. 196 pp. (Describes 85 species.)

Peattie, D. C. 1938. *A Prairie Grove*. New York: Literary Guild of America.

Reichman, O. J. 1987. *Konza Prairie: A Tallgrass Natural History*. Lawrence: University Press of Kansas.

Risser, P. G., E. C. Berry, H. D. Blocker, S. W. May, W. J. Parton, and J. A. Weins. 1981. *The True Prairie Ecosystem*. Stroudsbury, PA: Hutchinson Ross.

Samson, F. B., and F. L. Knopf. 1996. *Prairie Conservation: Preserving North America's Most Endangered Ecosystem*. Washington, DC: Island Press. 240 pp.

Samson, F. B., and W. R. Ostlie. 1998. Grasslands. Pp. 437–471, in *Status and Trends of the Nation's Biological Resources* (M. J. Mace et al., eds.). 2 vols. Reston, MD: US Department of the Interior, US Geological Survey.

Savage, C. 2004. *Prairie: A Natural History*. Vancouver, BC: Greystone Books. 320 pp.

Sears, P. 1935. *Deserts on the March*. Norman: University of Oklahoma Press. 260 pp. (Reprint, 1980).

———. 1969. *Lands beyond the Forest*. New York: Prentice Hall.

Shantz, H. L. 1923. The natural vegetation of the Great Plains region. *Annals of the Association of American Geographers* 13(2): 81–107.

Shantz, H. L., and R. Zon. 1924. Natural Vegetation. 28 pp. in: *Atlas of American Agriculture*, Washington, DC: US Department of Agriculture.

Sheviak, C. J., and M. L. Bowles. 1986. The prairie fringed orchids: A pollinator isolated pair. *Rhodora* 88: 167–190.

Sipes, S. D., and V. J. Tepedino. 1995. Reproductive biology of the rare orchid *Spiranthes diluvialis*: Breeding system, pollination, and implications for conservation. *Conservation Biology* 9: 929–938.

Smith, A. 1996. *Big Bluestem: Journey*

*into the Tall Grass*. Tulsa, OK: Council Oak Books. 304 pp.

Steiger, T. L. 1930. Structure of prairie vegetation. *Ecology* 11: 170–217.

Steinauer, G. 1998. In search of prairie orchids. *Nebraskaland* 76(2): 9.

Steinauer, G., and S. B. Rolfsmeier. 2003. *Terrestrial Natural Communities of Nebraska* (Version III). Lincoln: Nebraska Game and Parks Commission. 162 pp.

Trimble, D. E. 1980. *The Geologic Story of the Great Plains. Geological Survey Bulletin* 1493. Washington, DC: US Department of the Interior, US Geological Survey. 55 pp.

van der Cingel, N. A. 1995. *An Atlas of Orchid Pollination: European Orchids*. Rotterdam, Netherlands: A. A. Balkema.

van der Valk, A., ed. 1989. *Northern Prairie Wetlands*. Ames: Iowa State University Press.

Weaver, J. E. 1954. *North American Prairie*. Lincoln, NE: Johnsen.

———. 1965. *Native Vegetation of Nebraska*. Lincoln: University of Nebraska Press.

———. 1968. *Prairie Plants and Their Environment: A Fifty-Year Study in the Midwest*. Lincoln: University of Nebraska Press.

Weaver, J. E., and F. E. Clements. 1954. *Plant Ecology*. New York, NY: McGraw-Hill.

Weaver, J. E., and F. W. Albertson. 1956. *Grasslands of the Great Plains: Their Nature and Use*. Lincoln, NE: Johnsen. 395 pp.

Weaver, J. E., and T. J. Fitzpatrick. 1954. The prairie. *Ecological Monographs* 4: 109–295. (Reprint 1980, Prairie Plains Resource Institute, Aurora, Nebraska)

Wells, P. V. 1970. Postglacial vegetational history of the Great Plains. *Science* 167: 1574–1582.

Winkler, S. 1997. *The Heartland: Illinois, Iowa, Nebraska*. The Smithsonian Guides to Natural America. New York: Random House. 304 pp.

———. 2004. *Prairie: A North American Guide*. Iowa City: University of Iowa Press. 146 pp.

Wishart, D. J., ed. 2004. *Encyclopedia of the Great Plains*. Lincoln: University of Nebraska Press. 940 pp.

Wright, H. E., Jr. 1970. Vegetational history of the Great Plains. Pp. 157–172, in *Pleistocene and Recent Environments of the Central Great Plains* (W. Dort, Jr. and J. K. Jones, Jr., eds.). Lawrence: University Press of Kansas.

## Mammals

Armstrong, D. M. 1972. Distribution of mammals in Colorado. *Monograph of the Museum of Natural History, the University of Kansas* 3: 1–425.

Benedict, R. A., H. H. Genoways, and P. W. Freeman. 2000. Shifting distributional patterns of mammals in Nebraska. *Transactions of the Nebraska Academy of Sciences* 26: 55–84.

Benedict, R. A., P. W. Freeman, and H. H. Genoways. 1996. Prairie legacies—mammals. Pp. 149–166, in *Prairie Conservation: Preserving America's Most Endangered Ecosystem* (F. B. Samson and F. L. Knopf, eds.). Covelo, CA: Island Press.

Cadieux, C. 1983. *Coyotes: Predators and Survivors*. New York: Stone Wall Press.

Chapman, J. A., and G. A. Feldhamer, eds. 1982. *Wild Mammals of North America: Biology, Management, and Economics*. Baltimore, MD: Johns Hopkins University Press. 1147 pp.

Clark, T. W., and M. R. Stromberg. 1987. *Mammals in Wyoming.* Lawrence: University Press of Kansas. 324 pp. (Includes 117 species.)

Epperson, C. 1978. The biology of the bobcat in Nebraska. MS thesis, University of Nebraska–Lincoln. 129 pp.

Feldhamer, G. A., B. C. Thompson, and J. A. Chapman. 2003. *Wild Mammals of North America: Biology, Management, and Conservation.* Baltimore, MD: John Hopkins University Press. 1,216 pp.

Genoways, H. H., J. D. Hoffman, P. W. Freeman, K. Geloso, R. Benedict, and J. J. Huebshman. 2008. *Mammals of Nebraska: Checklist, Key, and Bibliography.* Lincoln: *Bulletin of the University of Nebraska State Museum* 23: 1–92.

Grady, W. 1994. *The Nature of Coyotes.* Vancouver, BC: Douglas & McIntyre.

Hall, E. R. 1965. Handbook of mammals of Kansas. *Miscellaneous Publication of the Museum of Natural History, the University of Kansas* 7: 1–303.

——. 1981. *The Mammals of North America.* 2 vols. New York: Wiley. 1181 pp.

Hart, F. M., and J. A. King. 1966. Distress vocalizations of young in two subspecies of *Peromyscus maniculatus. Journal of Mammalogy* 47: 287–293.

Higgins, K. E., E. D. Stukel, J. M. Goulet, and D. C. Backlund. 2002. *Wild Mammals of South Dakota.* Pierre: South Dakota Department of Game, Fish and Parks. 278 pp. (Includes 95 species.)

Jense, G. K., and R. L. Linder. 1970. Food habits of badgers in South Dakota. *Proceedings of the South Dakota Academy of Science* 49: 37–41.

Jones, J. Knox, Jr. 1964. *Distribution and Taxonomy of Mammals of Nebraska.* Lawrence: Museum of Natural History, University of Kansas Publication 16(1): 1–356.

Jones, J. Knox, Jr., D. N. Armstrong, and J. R. Choate. 1985. *Guide to Mammals of the Plains States.* Lincoln: University of Nebraska Press. 371 pp. (Coverage is from North Dakota to Oklahoma; includes 138 species.)

Jones, J. Knox, Jr., D. N. Armstrong, R. S. Hoffman, and C. Jones. 1983. *Guide to Mammals of the Northern Great Plains.* Lincoln: University of Nebraska Press. 371 pp. (Coverage is from North Dakota to Kansas; includes 105 species.)

Jones, J. Knox, Jr., and J. R. Choate. 1980. Annotated checklist of mammals of Nebraska. *Prairie Naturalist* 12(2): 43–53. http://digitalcommons.unl.edu/museummammalogy/164/

Kays, R. W., and D. E. Wilson. 2002. *Mammals of North America.* Princeton, NJ: Princeton University Press. (Includes 442 species.)

King, J. A., ed. 1967. *Biology of Peromyscus (Rodentia).* American Society of Mammalogists, Special Publication 110.

Leydet, F. 1988. *The Coyote: Defiant Songdog of the West.* Norman: University of Oklahoma Press.

Long, C. A. 1965. *The Mammals of Wyoming.* Lawrence: Museum of Natural History, University of Kansas 14: 493–758. (Includes 117 species.)

McCarley, H. 1966. Annual cycle, population dynamics, and adaptive behavior of *Citellus tridecemlineatus. Journal of Mammalogy* 47: 294–316.

Murie, J. O., and G. R. Michener, eds. 1984. *The Biology of Ground-dwelling Squirrels.* Lincoln: University of Nebraska Press.

Murie, O. J. 1954. *A Field Guide to Animal Tracks.* Boston: Houghton Mifflin. 400 pp.

Neal, E. G. 1996. *The Natural History of Badgers.* New York: Facts on File.

Pomerantz, S. M., and L. G. Clemens, 1981. Ultrasonic vocalizations in male deer mice (*Peromyscus maniculatus bairdi*): Their role in male sexual behavior. *Physiology and Behavior* 27: 869–872.

Reid, F. 2006. *Peterson Field Guide to Mammals of North America*. 4th ed. New York: Houghton Mifflin Harcourt. 608 pp.

Ryden, H. 1977. *God's Dog: A Celebration of the North American Coyote*. New York: Coward, McCann & Geoghegan.

Schwartz, C. W., and E. R. Schwartz. 2016. *The Wild Mammals of Missouri*. 3rd ed. Columbia: University of Missouri Press and Jefferson City: Missouri Department of Conservation. 396 pp. (Includes 67 species.)

Whitaker, J. O., Jr. 1996. *National Audubon Society Field Guide to North American Mammals*. 2nd ed. New York: Knopf.

Wilson, D. E., and S. Ruff. 1999. *The Smithsonian Book of North American Mammals*. Washington, DC: Smithsonian Institution Press. 710 pp.

### Birds

Ailes, I. W. 1980. Breeding biology and habitat use of the upland sandpiper in central Wisconsin. *Passenger Pigeon* 42: 53–63.

Ailes, I. W., and J. E. Toepfer. 1977. Home range and daily movement of radio-tagged upland sandpipers in central Wisconsin. *Inland Bird Banding News* 49: 203–212.

Alderfer, J., ed. 2017. *Complete Birds of North America*. 7th ed. Washington, DC: Smithsonian Institution Press. 664 pp. (The best field guide for advanced birders, with 1,023 species).

Anstey, D., S. K. Davis, D. C. Duncan, and M. Skeet. 1995. Distribution and habitat requirements of eight grassland songbird species in southern Saskatchewan. Regina: Saskatchewan Wetland Conservation Corporation. 11 pp.

Askins, R. A. 1993. Population trends in grassland, shrubland, and forest birds in eastern North America. *Current Ornithology* 11: 11–34.

———. 1999. History of grassland birds in eastern North America. Pp. 60–71, in *Ecology and Conservation of Grassland Birds of the Western Hemisphere* (P. D. Vickery and J. R. Herkert, eds.). Studies in Avian Biology No. 19. Camarillo, CA: Cooper Ornithological Society. 299 pp. https://sora.unm.edu/node/84

Askins, R. A., F. Chávez-Ramírez, B. C. Dale, C. A. Haas, J. R. Herkert, F. L. Knopf, and P. D. Vickery. 2007. *Conservation of Grassland Birds in North America: Understanding Ecological Processes in Different Regions: Report of the AOU Committee on Conservation*. Ornithological Monographs No. 64. Washington, DC: American Ornithologists' Union. 46 pp.

Aweida, M. K. 1995. Repertoires, territory size, and mate attraction in western meadowlarks. *Condor* 97: 1080–1083.

Baichich, P. J., and C. J. O. Harrison. 1997. *A Guide to the Nests, Eggs, and Nestlings of North American Birds*. 2nd ed. San Diego, CA: Academic Press. 348 pp.

Bakker, A., K. D. Withrow, and N. S. Thompson. 1983. Levels of organization in the song of the bobolink (Icteridae: Dolichonyxinae). *Zeitschrift für Tierpsychologie* [*Ethology*] 62: 105–114.

Basili, G. D. 1997. Continent-scale ecology and conservation of dickcissels. PhD diss., University of Wisconsin–Madison.

Beason, R. C. 1995. Horned lark (*Eremophila alpestris*). In *The Birds of North America*, no. 195 (A. Poole and F. Gill, eds.). Philadelphia: Academy of Natural Sciences, and Washington, DC: American Ornithologists' Union.

Beason, R. C., and E. C. Franks. 1974. Breeding behavior of the horned lark. *Auk* 91: 65–74.

Bent, A. C. [O. L. Austin, ed.] 1968. *Life Histories of North American Cardinals, Grosbeaks, Buntings, Towhees, Finches, Sparrows, and Allies*. In three parts. Washington, DC: Smithsonian Institution Press. 1889 pp. (Pt. 1: Cardinalidae, Fringillidae (part). Pt. 2: Emberizidae (part). Pt. 3: Emberizidae (part), Calcariidae).

Berry, G. A. 1971. The nesting biology of the dickcissel in north central Oklahoma. MS thesis, Oklahoma State University, Stillwater. 27 pp.

Blankespoor, G. W. 1970. The significance of nest and nest site microclimate for the dickcissel, *Spiza americana*. PhD diss., Kansas State University, Manhattan. 184 pp.

Bollinger, E. K. 1988. Breeding dispersion and reproductive success of bobolinks in an agricultural landscape. PhD diss., Cornell University, Ithaca, NY.

Borror, D. J. 1971. Songs of *Ammophila* sparrows occurring in the United States. *Wilson Bulletin* 83: 132–151.

Borror, D. J., and C. R. Reese. 1954. Analytical studies of Henslow's sparrow songs. *Wilson Bulletin* 66: 243–252.

Bowen, D. E., Jr. 1976. Coloniality, reproductive success, and habitat interactions in upland sandpipers (*Bartramia longicauda*). PhD diss., Kansas State University, Manhattan.

Boyd, R. L. 1976. Behavioral biology and energy expenditure in a horned lark population. PhD diss., Colorado State University, Fort Collins.

Brennan, L. A., and W. P. Kuvlesky, Jr. 2005. North American grassland birds: An unfolding conservation crisis? *Journal of Wildlife Management* 69: 1–13.

Clawson, R. L. 1991. Henslow's sparrow habitat site fidelity and reproduction in Missouri. Final report, Federal Aid Project No. W-14-R-45. Jefferson City: Missouri Department of Conservation. 18 pp.

Cody, M. L. 1968. On the methods of resource division in grassland bird communities. *American Naturalist* 102: 107–147. (Includes western meadowlark, Sprague's pipit, bobolink, horned lark, and the clay-colored, vesper, Savannah, grasshopper, and Baird's sparrows.)

D'Agincourt, L. G., and J. B. Falls. 1983. Variation of repertoire use in the eastern meadowlark, *Sturnella magna*. *Canadian Journal of Zoology* 61: 1086–1093.

Dale, B. C. 1983. Habitat relationships of seven species of passerine birds at Last Mountain Lake, Saskatchewan. MS thesis, University of Regina, Regina, SK. 119 pp.

———. 1984. Birds of grazed and ungrazed grasslands in Saskatchewan. *Blue Jay* 42: 102–104. (Includes horned lark, western meadowlark, Sprague's pipit, chestnut-collared longspur, and the clay-colored, vesper, Savannah, and Baird's sparrows.)

Dale, B. C., P. A. Martin, and P. S. Taylor. 1997. Effects of hay management on grassland songbirds in Saskatchewan. *Wildlife Society Bulletin* 25: 616–626.

Darwin, C. 1871. *The Descent of Man, and Selection in Relation to Sex*. London: John Murray.

Davis, S. K., and S. G. Sealy. 2000. Cowbird parasitism and predation in grassland fragments of southwestern Manitoba. Pp. 220–228, in *The*

*Biology and Management of Cowbirds and Their Hosts* (J. N. M. Smith, T. L. Cook, S. I. Rothstein, S. K. Robinson, and S. G. Sealy, eds.). Austin: University of Texas Press.

Dechant, D. J., M. L. Sondreal, D. H. Johnson, L. D. Igl, C. M. Goldade, M. P. Nenneman, and B. R. Euliss. 1999. *Effects of Management Practices on Grassland Birds*. Jamestown, ND: Northern Prairie Wildlife Research Center. (The accounts that relate to Spring Creek Prairie and eastern Nebraska tallgrass prairie breeding birds are those for the dickcissel, field sparrow, grasshopper sparrow, and Henslow's sparrow.)

Falls, J. B., A. G. Horn, and T. E. Dickinson. 1988. How western meadowlarks classify their songs: Evidence from song matching. *Animal Behavior* 36: 579–585.

Fink, E. J. 1983. Male behavior, territory quality, and female choice in the dickcissel (*Spiza americana*). PhD diss., Kansas State University, Manhattan. 79 pp.

Fleischer, R. C. 1986. Brood parasitism of brown-headed cowbirds in a simple host community in eastern Kansas. *Kansas Ornithological Society Bulletin* 37: 21–29.

Friedmann, H. 1963. *Host Relations of the Parasitic Cowbirds*. United States National Museum Bulletin 233. Washington DC: Smithsonian Institution (Museum of Natural History). 276 pp.

Graber, R. R., and J. W. Graber. 1963. A comparative study of bird populations in Illinois, 1906–1909 and 1956–1958. *Illinois Natural History Survey Bulletin* 28(3). Champaign: Illlinois Natural History Survey. 145 pp. (Includes upland sandpiper, horned lark, bobolink, eastern and western meadowlarks, dickcissel, and the Savannah, grasshopper, vesper, and lark sparrows.)

Hands, H. M., R. D. Drobney, and M. R. Ryan. 1989. Status of the Henslow's sparrow in the northcentral United States. Columbia, MO: US Fish & Wildlife Service, Missouri Cooperative Fish and Wildlife Research Unit. 12 pp.

Harmeson, J. P. 1974. Breeding ecology of the dickcissel. *Auk* 91: 348–359.

Hatch, S. A. 1983. Nestling growth relationships of brown-headed cowbirds and dickcissels. *Wilson Bulletin* 95: 669–671.

Hergenrader, G. L. 1962. The incidence of nest parasitism by the brown-headed cowbird (*Molothrus ater*) on roadside nesting birds in Nebraska. *Auk* 79: 85–88.

Herkert, J. R. 1991. An ecological study of the breeding birds of grassland habitats within Illinois. PhD diss., University of Illinois, Urbana. (Includes upland sandpiper; eastern meadowlark; grasshopper, Savannah, and Henslow's sparrows; and dickcissel.)

———. 1994. The effect of habitat fragmentation on Midwestern grassland bird communities. *Ecological Applications*. 4: 461–471.

Herkert, J. R., and W. D. Glass. 1999. Henslow's sparrow response to prescribed fire in an Illinois prairie remnant. Pp. 160–164, in *Ecology and Conservation of Grassland Birds of the Western Hemisphere* (P. D. Vickery and J. R. Herkert, eds.). Studies in Avian Biology No. 19. Camarillo, CA: Cooper Ornithological Society. 299 pp. https://sora.unm.edu/node/84

Higgins, K. F., and L. M. Kirsh. 1975. Some aspects of the breeding biology of the upland sandpiper in North Dakota. *Wilson Bulletin* 87: 96–101.

Higgins, K. F., H. R. Duebbert, and R. B. Oetting. 1969. Nesting of the upland plover on the Missouri Coteau. *Prairie Naturalist* 1: 45–48.

Hill, R. A. 1976. Host-parasite relationships of the brown-headed cowbird in a prairie habitat of west-central Kansas. *Wilson Bulletin* 88: 555–565.

Horn, A. G., and J. B. Falls. 1988. Structure of western meadowlark (*Sturnella neglecta*) song repertoires. *Canadian Journal of Zoology* 66: 284–288.

——. 1991. Song switching in mate attraction and territory defense by western meadowlarks (*Sturnella neglecta*). *Ethology* 87: 262–268.

Horn, A. G., T. E. Dickinson, and J. B. Falls. 1993. Male quality and song repertoires in western meadowlarks (*Sturnella neglecta*). *Canadian Journal of Zoology* 71: 1059–1061.

Houston, C. S., and D. E. Bowen, Jr. 2001. Upland sandpiper (*Bartramia longicauda*). In *The Birds of North America*, no. 580. (A. Poole and F. Gill, eds.). Philadelphia: Academy of Natural Sciences, and Washington, DC: American Ornithologists' Union.

Hyde, A. S. 1939. The life history of Henslow's sparrow, *Passerherbulus henslowii* (Audubon). Miscellaneous Publications No. 41, Museum of Zoology, University of Michigan. Ann Arbor: University of Michigan Press. https://deepblue.lib.umich.edu/handle/2027.42/56286

Johnsgard, P. A. 1973. *Grouse and Quails of North America*. Lincoln: University of Nebraska Press. 553 pp. http://digitalcommons.unl.edu/bioscigrouse/1/

——. 1979. *Birds of the Great Plains: Breeding Species and Their Distribution*. Lincoln: University of Nebraska Press. 538 pp.

——. 1981. *The Plovers, Sandpipers, and Snipes of the World*. Lincoln: University of Nebraska Press. 492 pp.

——. 1988. *North American Owls: Biology and Natural History*. Washington, DC: Smithsonian Institution Press. 295 pp.

——. 1990. *Hawks, Eagles, and Falcons of North America: Biology and Natural History*. Washington, DC: Smithsonian Institution Press. 403 pp.

——. 1994. *Arena Birds: Sexual Selection and Behavior*. Washington, DC: Smithsonian Institution Press. 330 pp.

——. 1997. *The Avian Brood Parasites: Deception at the Nest*. New York: Oxford University Press. 409 pp.

——. 2001a. *Prairie Birds: Fragile Splendor in the Great Plains*. Lawrence: University Press of Kansas. 331 pp.

——. 2001c. Historic birds of Lincoln's Salt Basin Wetlands and Nine-mile Prairie. *Nebraska Bird Review* 68: 132–136. https://digitalcommons.unl.edu/nebbirdrev/30/

——. 2002. *Grassland Grouse and Their Conservation*. Washington, DC: Smithsonian Institution Press. 157 pp.

——. 2005. *A Nebraska Bird-Finding Guide*. University of Nebraska–Lincoln DigitalCommons and Zea Books. 152 pp. https://digitalcommons.unl.edu/zeabook/5/

——. 2010. The drums of April. *Prairie Fire*, April, pp. 12–13. http://www.prairiefirenewspaper.com/2010/04/the-drums-of-april (Prairie grouse display)

——. 2012d. *Wings over the Great Plains: Bird Migrations in the Central Flyway*. University of Nebraska–Lincoln DigitalCommons and Zea Books. 249 pp. http://digitalcommons.unl.edu/zeabook/13/

——. 2016. *The North American Grouse: Biology and Behavior*. University of Nebraska–Lincoln DigitalCommons and Zea Books. 183 pp. http://digitalcommons.unl.edu/zeabook/41/

——. 2017. *The North American Quails,*

*Partridges, and Pheasants.* University of Nebraska–Lincoln DigitalCommons and Zea Books. 131 pp. http://digitalcommons.unl.edu/zeabook/58/

———. 2018b. *The Birds of Nebraska.* Rev. ed. University of Nebraska–Lincoln DigitalCommons and Zea Books. http://digitalcommons.unl.edu/zeabook/65/

Johnson, D. H. 1972–74. Breeding bird populations of selected grasslands in east-central North Dakota. *American Birds* 26: 970–975; 27: 989–990; 28: 1030–1031.

Johnson, R. G., and S. A. Temple. 1986. Assessing habitat quality for birds nesting in fragmented tallgrass prairies. Pp. 245–249, in *Modeling Habitat Relationships of Terrestrial Vertebrates* (J. Verner, M. L. Morrison, and C. J. Ralph, eds.). Madison: University of Wisconsin Press.

Jones, S. L., J. S. Dieni, and P. J. Gouse. 2010. Reproductive biology of a grassland songbird community in north-central Montana. *Wilson Journal of Ornithology* 122: 455–464.

Kaufman, K., and R. Bowers. 2005. *Kaufman Field Guide to Birds of North America.* Boston: Houghton Mifflin Harcourt. 392 pp. (A popular field guide, especially for starting birders, that uses digital photos.)

Kinstler, K. A., and T. A. Sordahl. 1994. A comparison of perch use by vocalizing eastern and western meadowlarks. *Prairie Naturalist* 26: 195–200.

Kirsch, L., and K. F. Higgins. 1976. Upland sandpiper nesting and management in North Dakota. *Wildlife Society Bulletin* 4: 16–20.

Kirsch, L. M., H. F. Duebbert, and A. D. Kruse. 1978. Grazing and haying effects on habitats of upland nesting birds. *Transactions of the North American Wildlife Natural Resources Conference* 43: 486–497.

Klippenstine, D. R., and S. G. Sealy. 2008. Differential ejection of cowbird eggs and non-mimetic eggs by grassland passerines. *Wilson Journal of Ornithology* 120: 667–673.

Knapton, R. W. 1988. Nesting success is higher for polygynously mated females than for monogamously mated females in the eastern meadowlark. *Auk* 105: 325–329.

Knight, S. 1995. Cowbird parasitism, nest predation, and host selection in fragmented grasslands of southwestern Manitoba. PhD diss., University of Manitoba, Winnipeg.

Lanyon, W. E. 1994. Western meadowlark (*Sturnella neglecta*). In *The Birds of North America*, no. 104. (A. Poole and F. Gill, eds.). Philadelphia: Academy of Natural Sciences, and Washington, DC: American Ornithologists' Union.

———. 1995. Eastern meadowlark (*Sturnella magna*). In *The Birds of North America*, no. 160. (A. Poole and F. Gill, eds.). Philadelphia: Academy of Natural Sciences, and Washington, DC: American Ornithologists' Union.

Long, C. A., C. F. Long, J. Knops, and D. H. Matulionis. 1965. Reproduction in the dickcissel. *Wilson Bulletin* 77: 251–256.

Lowther, P. E. 1993. Brown-headed cowbird (*Molothrus ater*). In *The Birds of North America*, no. 47. (A. Poole and F. Gill, eds.). Philadelphia: Academy of Natural Sciences, and Washington, DC: American Ornithologists' Union.

Ludlow, S. R., Mark Brigham, and S. K. Davis. 2014. Nesting ecology of grassland songbirds: Effects of predation, parasitism, and weather. *Wilson Journal of Ornithology* 126: 686–689.

Martin, S. G. 1967. Breeding biology of the bobolink. MS thesis, University of Wisconsin–Madison.

Martin, S. G., and T. A. Gavin. 1995. Bobolink (*Dolichonyx oryzivorus*). In *The Birds of North America*, no. 176. (A. Poole and F. Gill, eds.). Philadelphia: Academy of Natural Sciences, and Washington, DC: American Ornithologists' Union.

Mineau, P., and M. Whiteside. 2013. Pesticide acute toxicity is a better correlate of US grassland bird declines than agricultural intensification. *PLOS ONE* 8(2): e57457. https://doi.org/10.1371/journal.pone.0057457

Mollhoff, W. J. 2016. *The Second Nebraska Breeding Bird Atlas*. Bulletin of the University of Nebraska State Museum 29. Lincoln: University of Nebraska State Museum. 304 pp. (Includes 225 species.)

Newman, O. A. 1970. Cowbird parasitism and nesting success of lark sparrows in southern Oklahoma. *Wilson Bulletin* 82: 304–309.

Ortega, C. 1998. *Cowbirds and Other Brood Parasites*. Tucson: University of Arizona Press. 371 pp.

Peer, B. D., S. K. Robinson, and J. R. Herkert. 2000. Egg rejection by cowbird hosts in grasslands. *Auk* 117: 892–901.

Peterjohn, B. G., and J. R. Sauer. 1999. Population status of North American grassland birds from the North American Breeding Bird Survey, 1966–1996. Pp. 27–44, in *Ecology and Conservation of Grassland Birds of the Western Hemisphere* (P. D. Vickery and J. R. Herkert, eds.). Studies in Avian Biology No. 19. Cooper Ornithological Society. 299 pp. https://sora.unm.edu/node/84

Peterson, R. T. 2002. *Peterson Field Guide to Birds of Eastern and Central North America*. Boston: Houghton Mifflin. 524 pp. (A popular and classic field guide that uses paintings, excellent for beginning birders.)

Rich, T. D., C. J. Beardmore, H. Berlanga, P. J. Blancher, M. S. W. Bradstreet, G. S. Butcher, D. W. Demarest, E. H. Dunn, W. C. Hunter, E. E. Iñigo-Elias, J A. Kennedy, A. M. Martell, A. O. Panjabi, D. N. Pashley, K. V. Rosenburg, C. M. Rustay, J. S. Wendt, and T. C. Will. 2004. *North American Landbird Conservation Plan*. Ithaca, NY: Partners in Flight and Cornell University Laboratory of Ornithology. 84 pp. https://www.partnersinflight.org/resources/north-american-landbird-conservation-plan/

Robins, J. D. 1971a. A study of Henslow's sparrow in Michigan. *Wilson Bulletin* 83: 39–48.

———. 1971b. Differential niche utilization in a grassland sparrow. *Ecology* 52: 1065–1070.

Rodewald, P., ed. 2015. *The Birds of North America Online*. Ithaca, NY: Cornell University Laboratory of Ornithology. https://birdsna.org (Modern life histories of more than 600 species of North American bird species, mostly updated from out-of-print versions of 1990s and early 2000s accounts.)

Rotenberry, J. T., and W. D. Klimstra. 1970. The nesting ecology and reproductive performance of the eastern meadowlark. *Wilson Bulletin* 82: 243–267.

Ryan, M. R., L. W. Burger, Jr., D. P. Jones, and A. P. Wywialowski. 1998. Breeding ecology of greater prairie-chickens (*Tympanuchus cupido*) in relation to prairie landscape configuration. *American Midland Naturalist* 140: 111–121.

Salt, W. R. 1966. A nesting study of *Spizella pallida*. *Auk* 83: 274–281.

Sample, D. W. 1989. Grassland birds in southern Wisconsin: Habitat preferences, population trends, and response to land-use changes. MS thesis, University of Wisconsin–Madison. 588 pp.

Sample, D. W., and M. J. Mossman. 1997. *Managing Habitat for Grassland Birds: A Guide for Wisconsin*. Madison: Wisconsin Department of Natural Resources. (Includes upland sandpiper, horned lark, western and eastern meadowlarks, bobolink, dickcissel, vesper sparrow, and grasshopper sparrow.)

Sauer, J. R., J. E. Hines, I. Thomas, J. Fallon, and G. Gough. 1997.The North American Breeding Bird Survey results and analysis, 1966–1998. Version 96.4. Laurel, MD: USGS Patuxent Wildlife Research Center.

Sauer. J. R., W. A. Link, J. E. Fallon, K. L. Pardieck, and D. J. Ziolkowski, Jr. 2013. The North American Breeding Bird Survey 1966–2011: Summary analysis and species accounts. *North American Fauna* 79: 1–32.

Sauer, J. R., D. K. Neevin, J. E. Hines, D. J. Ziolkowski, Jr., K. L. Pardieck, J. E. Fallon, and W. A. Link. 2017. The North American Breeding Bird Survey results and analysis. 1966–1998. Version 2.07. Laurel, MD: USGS Patuxent Wildlife Research Center.

Sharpe, R. S., W. R. Silcock, and J. G. Jorgensen. 2001. *Birds of Nebraska: Their Distribution and Temporal Occurrence*. Lincoln: University of Nebraska Press. (See also *Birds of Nebraska Online*, https://birds.outdoornebraska.gov/)

Short, L. L., Jr. 1965. Hybridization in the flickers (*Colaptes*) of North America. *Bulletin of the American Museum of Natural History* 129: 311–428.

Sibley, C. G., and D. A. West. 1959. Hybridization in the rufous-sided towhees of the Great Plains. *Auk* 76: 326–338.

Sibley, D. A. 2003. *The Sibley Field Guide to the Birds of Western North America*. New York: Knopf. (A popular field guide that uses paintings.)

Smith, C. R. 1992. Henslow's sparrow, *Ammodramus henslowi*. Pp. 315–330, in *Migratory Nongame Birds of Management Concern in the Northeast* (K. J. Schneider and D. M. Pence, eds.). Newton Corner, MA: US Fish & Wildlife Service. 400 pp.

Smith, J. N. M., T. L. Cook, S. I. Rothstein, S. K. Robinson, and S. G. Sealy, eds. 2000. *Ecology and Management of Cowbirds and Their Hosts*. Austin: University of Texas Press. 496 pp.

Steigman, K. L. 1993. Nesting ecology of the dickcissel (*Spiza americana*) in a tallgrass prairie in north central Texas. PhD diss., University of North Texas, Denton. 133 pp.

Swanson, D. A. 1996. Nesting ecology and nesting habitat requirements of Ohio's grassland-nesting birds: A review. Ohio Fish and Wildlife Report 13. Columbus: Ohio Division of Wildlife, Department of Natural Resources. 60 pp. (Includes upland sandpiper, bobolink, eastern and western meadowlarks, dickcissel, and the grasshopper, Henslow's, vesper, and lark sparrows.)

US Fish & Wildlife Service (USFWS). 2008. *Birds of Conservation Concern 2008*. US Fish & Wildlife Service, Division of Migratory Bird Management, Arlington, VA. 85 pp. https://www.fws.gov/migratorybirds/pdf/grants/BirdsofConservationConcern2008.pdf

Vance, J., and N. Paothong. 2012. *Save the Last Dance: A Story of North American Grassland Grouse*. Columbia MO: Noppadol Paothong Photography.

Vickery, P. D. 1996. Grasshopper sparrow (*Ammodramus savannarum*). In *The Birds of North America*, no. 239. (A. Poole and F. Gill, eds.). Philadelphia: Academy of Natural Sciences, and Washington, DC: American Ornithologists' Union.

Vickery, P. D., and J. R. Herkert, eds. 1999. *Ecology and Conservation of Grassland Birds of the Western Hemisphere*. Studies in Avian Biology No. 19. Cooper Ornithological Society. 299 pp. https://sora.unm.edu/node/84

Vickery, P. D., M. L. Hunter, and S. M. Scott. 1994. Effects of habitat area on the distribution of grassland birds in Maine. *Conservation Biology* 8: 1087–1097. (Includes upland sandpiper, bobolink, eastern meadowlark, grasshopper sparrow, and vesper sparrow.)

Wells, J. V., and P. D. Vickery. 1994. Extended flight-songs of vesper sparrows. *Wilson Bulletin* 106: 696–702.

Whelan, D. B. 1940. Birds of a surviving area of original prairie land in eastern Nebraska. *Nebraska Bird Review* 8(2): 50–55. http://digitalcommons.unl.edu/nebbirdrev/882/

Winter, M. 1999. Nesting biology of dickcissels and Henslow's sparrows in southwestern Missouri prairie fragments. *Wilson Bulletin* 111: 515–525.

Zimmerman, J. L. 1993. *The Birds of Konza*. Lawrence: University Press of Kansas. 186 pp. (Includes 208 species.)

## Amphibians and Reptiles

Anderson, P. 1965. *The Reptiles of Missouri*. Columbia: University of Missouri Press. (Updated by Johnson, 2000.)

Ballinger, R. E., J. D. Lynch, and G. R. Smith. 2010. *Amphibians and Reptiles of Nebraska*. Oro Valley, AZ: Rusty Lizard Press, and Lincoln: University of Nebraska Press. 400 pp. (Includes 13 amphibians and 48 reptiles plus 151 distribution maps and diagrams.)

Ballinger, R. E., J. D. Lynch, and P. H. Cole. 1979. Distribution and natural history of amphibians and reptiles in western Nebraska, with ecological notes on the herpetiles of Arapaho Prairie. *Prairie Naturalist* 11: 65–74.

Baxter, G. T., and M. D. Stone. 2011. *Amphibians and Reptiles of Wyoming*. 2nd ed. Cheyenne: Wyoming Naturalist. 192 pp. (Covers all 42 species of Wyoming herpetiles, with 280 color photos.)

Behler, J. L., and F. W. King. 1996. *The Audubon Society Field Guide to North American Reptiles and Amphibians*. New York: Knopf. 743 pp. (Coverage includes 283 reptiles and 194 amphibians, illustrated with 657 color photos.)

Benedict, R. 1996. Snappers, soft-shells, and stinkpots: The turtles of Nebraska. *Museum Notes* Lincoln: University of Nebraska State Museum 96: 1–4. (Describes the nine Nebraska turtle species.)

Caldwell, J. P., and J. T. Collins. 1981. *Turtles in Kansas*. Lawrence, KS: AMS Publishing.

Carpenter, C. C. 1960. Aggressive behavior and social dominance in the six-lined racerunner. *Animal Behavior* 8: 61–66.

Collins, J. T. 1993. *Amphibians and Reptiles of Kansas*. Lawrence: University of Kansas Museum of Natural History Public Education Series 13. 397 pp. (Includes 96 color photographs of all the Kansas herpetile species.)

Conant, R., and J. T. Collins. 1998. *Reptiles and Amphibians of Eastern and Central North America*. Boston: Houghton Mifflin. 640 pp. (See Powell, Conant, and Collins, 2016 for new edition.)

Duellman, W. E., and L. Trueb. 1986. *Biology of Amphibians*. New York: McGraw-Hill. 670 pp.

Ernst, C. H., J. E. Lovich, and R. W. Barbour. 1994. *Turtles of the United States and Canada*. Washington, DC: Smithsonian Institution Press. 578 pp. (Biology of the 56 turtles of the United States and Canada.)

Farrar, J. 1998. Box turtles: Life in the fast lane. *NEBRASKAland* 76(5): 24–33. (Ornate box turtle)

Ferraro, D. n.d. *A Guide to Snakes, Turtles, Frogs, Lizards, and Salamanders*. http://snr.unl.edu/herpneb/ (Identification guide to the Nebraska species.)

Fitch, H. S. 1954. Life history and ecology of the five-lined skink, *Eumeces fasciatus*. Lawrence: University of Kansas Museum of Natural History Publications 8: 2–156.

———. 1968. Natural history of the six-lined racerunner (*Cnemidophorus sexlineatus*). Lawrence: University of Kansas Museum of Natural History Publications 11: 11–62.

Fogell, D. D. 2010. *A Field Guide to the Amphibians and Reptiles of Nebraska*. Lincoln: University of Nebraska Conservation and Survey Division. 158 pp. (Includes 14 amphibians and 48 reptiles.)

Hammerson, G. A. 1999. *Amphibians and Reptiles in Colorado*. 2nd ed. Boulder: University of Colorado Press. 484 pp.

Holycross, A. T. 1995. Movements and natural history of the prairie rattle snake (*Crotalis viridis viridis*) in the Sandhills of Nebraska. MS thesis, University of Nebraska at Omaha. 88 pp.

Hudson, G. E. 1942. *The Amphibians and Reptiles of Nebraska*. Lincoln: University of Nebraska Conservation and Survey Division, Bulletin 22. 146 pp. (See also Ballinger, Lynch, and Smith, 2010.)

Iverson, J. B. 1975, 1977. Notes on Nebraska reptiles. *Transactions of the Kansas Academy of Science* 78: 51–62, 80: 55–59.

Iverson, J. B., and G. R. Smith. 1993. Reproductive ecology of the painted turtle (*Chrysemys picta*) in the Nebraska Sandhills and across its range. *Copeia* 1993(1): 1–21.

Johnson, T. R. 2000. *The Amphibians and Reptiles of Missouri*. Rev. ed. Jefferson City: Missouri Department of Conservation. 400 pp. (Includes color photos and biology of the Missouri species.)

Jones, S. M., and D. L. Droge. 1980. Home range size and special distribution of two sympatric lizard species (*Sceloperus undulatus* and *Holbrookia maculata*) in the Sandhills of Nebraska. *Herpetologica* 36: 127–132.

Joy, J. E., and D. Crews. 1985. Social dynamics of group courtship behavior in male red-sided garter snake (*Thamnophis sirtalis parietalis*). *Journal of Comparative Psychology* 99: 145–49.

Kardong, K. V. 1980. Gopher snakes and rattlesnakes: Presumptive Batesian mimicry. *Northwest Science* 54: 1–4.

Kiesow, A. M. 2006. *Field Guide to Amphibians and Reptiles of South Dakota*. Pierre: South Dakota Department of Game, Fish, and Parks. 178 pp.

Klauber, L. M. 1972. *Rattlesnakes: Their Habits, Life Histories, and Influence on Mankind*. 2nd ed. 2 vols. Berkeley: University of California Press.

Krupa, J. 1994. Breeding biology of the Great Plains toad in Oklahoma. *Journal of Herpetology* 28: 217–224.

Kruse, K. C. 1981. Phonotactic responses of female northern leopard frogs (*Rana pipiens*) to *Rana blairi*, a presumed hybrid, and conspecific mating trills. *Journal of Herpetology* 13: 145–150.

Legler, J. M. 1960. Natural history of the ornate box turtle, *Terrapene ornata*. Lawrence: University of Kansas Museum of Natural History Publications 11: 527–660.

Littlejohn, M. J., and R. S. Oldham. 1968. *Rana pipiens* complex mating call structure and taxonomy. *Science* 162: 1003–1995.

Lynch, J. D. 1985. Annotated checklist of the amphibians and reptiles of Nebraska. *Transactions of the Nebraska Academy of Sciences* 13: 33–57.

Oliver, J. A. 1955. *The Natural History of North American Amphibians and Reptiles*. Princeton, NJ: Van Nostrand.

Powell, R., R. Conant, and J. T. Collins. 2016. *Field Guide to Reptiles and Amphibians of Eastern and Central North America*. 4th ed. Boston: Houghton Mifflin. 474 pp. (Coverage extends to western border of Nebraska and includes 46 pages of paintings, 202 color photos, and numerous drawings of 501 species.)

Rossman, D. A., N. B. Ford, and R. A. Seigel. 1996. *The Garter Snakes: Evolution and Ecology*. Norman: University of Oklahoma Press.

Smith, H. M. 1946. *Handbook of Lizards: Lizards of the United States and Canada*. Ithaca, NY: Cornell University Press. 557 pp. (Includes 136 species.)

———. 1956. Handbook of amphibians and reptiles of Kansas. Lawrence: University of Kansas Museum of Natural History Miscellaneous Publications 9: 1–36. (Updated by Collins, 1993.)

Smith, H. M., and E. D. Brodie, Jr. 1983. *Reptiles of North America: A Guide to Field Identification*. New York: St. Martin's Press. 240 pp. (Includes 278 species.)

Stebbins, R. C. 1985. *A Field Guide to Western Reptiles and Amphibians*. 2nd ed. Boston: Houghton Mifflin.

336 pp. (Includes 244 species.)

Stebbens, R. C., and N. W. Cohen. 1995. *A Natural History of Amphibians*. Princeton, NJ: Princeton University Press. 316 pp.

Werner, J. K., B. A. Maxwell, P. Hendricks, and D. L. Flath. 2004. *Amphibians and Reptiles of Montana*. Missoula, MT: Mountain Press. 262 pp. (Covers 32 species of amphibians and reptiles of Montana.)

Wheeler, G. C., and J. Wheeler. 1966. *The Amphibians and Reptiles of North Dakota*. Grand Forks: University of North Dakota Press. 104 pp. (Covers about 28 species of North Dakota herpetiles.)

Wright, A. H., and A. A. Wright. 1949. *Handbook of the Frogs and Toads of the United States and Canada*. Ithaca, NY: Cornell University Press. 248 pp. (Outdated but useful nationwide coverage.)

———. Wright. 1957. *Handbook of the Snakes of the United States and Canada*. Ithaca, NY: Cornell University Press. 554 pp. (Outdated but useful nationwide coverage.)

Zim, H. S., and H. M. Smith. 2001. *Reptiles and Amphibians*. New York: Macmillan. 160 pp. (Includes 212 species.)

### Insects (General)

Borror, D. C., and R. E. White. 1979. *A Field Guide to Insects: America North of Mexico*. Boston: Houghton Mifflin. 404 pp. (Includes 1,300 drawings and 142 color paintings; oriented to more taxonomically competent biologists than most readers but highly informative, especially as to family-level taxonomy.)

Eaton, E. R., and K. Kaufman. 2007. *Kaufman Field Guide to Insects of North America*. Boston: Houghton

Mifflin. 382 pp. (Oriented to nonprofessional biologists, with 2,350 color digital images; easily used but little information on taxonomy, habitats, or life histories.)

Golick, D., and M. Ellis. 2000. Bumble boosters. Lincoln: University of Nebraska Cooperative Extension EC 00-1564-S. 64 pp. (Pictorial key to all 19 species of Nebraska bumblebees.)

Milne, L., and M. Milne. 1981. *National Audubon Society Field Guide to North American Insects and Spiders*. New York: Knopf. 990 pp. (Includes 702 photos, covering more than 600 species. Good ecology and life history information. Intermediate between Borror and White [1979] and Eaton and Kaufman [2007] guides in technical approach, and perhaps the best of the too-few available national field guides to insects.)

Ratcliffe, B. C. 1991a. Scarab beetles. *NEBRASKAland* 66(4): 30–36. (Brief overview of Nebraska species.)

———. 1991b. *The Scarab Beetles of Nebraska*. Lincoln: University of Nebraska State Museum Bulletin 12. 333 pp. (A technical monograph of 197 species, with many illustrations and eight color paintings.)

Salsbury, G. A., and S. C. White. 2000. *Insects in Kansas*. Manhattan: Kansas Department of Agriculture. 523 pp. (More than 850 species are shown in color photos and briefly described, with an emphasis on economically important groups, plus coil-bound for flat page layout.)

Stawell, R. 1921. *Fabre's Book of Insects*. New York: Tudor. 271 pp. (Fascinating accounts of insect behavior by a famous nineteenth-century naturalist.)

Williams, P. H., R. W. Thorp, and L. L. Richardson. 2014. *Bumble Bees of North America: An Identification Guide*. Princeton, NJ: Princeton University Press. 208 pp. (Includes all 46 North American species, with color photos, maps, and diagrams.)

Wilson, J., and O. J. Messinger. 2015. *The Bees in Your Backyard: A Guide to North America's Bees*. Princeton, NJ: Princeton University Press. 288 pp. (Includes 900 color photos and is a comprehensive introduction to the 4,000 species of North American bees.)

### Butterflies and Moths

Beadle, D., and S. Leckie. 2012. *Peterson Field Guide to Moths of Northeastern North America* (Peterson Field Guides). Boston: Houghton Mifflin. 624 pp. (Uses digital color images to illustrate about 1,500 species.)

Brock, J. P., and K. Kaufman. 2006. *Kaufman Field Guide to Butterflies of North America*. Boston: Houghton Mifflin. 392 pp. (Probably the most useful for general readers among the many available butterfly field guides, this book uses 2,300 digital images to illustrate all butterfly species north of Mexico. The 2006 edition added four new plates to the original 2003 version.)

Covell, C. V. 1984. *A Field Guide to Moths of Eastern North America*. Boston: Houghton Mifflin, and Martinsville: Virginia Museum of Natural History. 528 pp. (Includes color photos of 1,500 species of moths; a second edition was published in 2005.)

Dankert, N., M. L. Brust, H. Nagel, and S. M. Spomer. 2005. *Butterflies of Nebraska*. Department of Biology, University of Nebraska–Kearney. http://www.lopers.net/student_org/NebraskaInverts/butterflies/home.htm (Version 5APR2005).

Dankert, N. E., and H. C. Nagel. 1988. Butterflies of the Niobrara Valley Preserve, Nebraska. *Transactions of the Nebraska Academy of Sciences* 16: 17–30. (Includes 70 species.)

Douglas, M. M. 1986. *The Lives of Butterflies*. Ann Arbor: University of Michigan Press.

Farrar, J. 1993. The making of a monarch. *NEBRASKAland* 71(6): 34–41.

Feltwell, J. 1986. *The Natural History of Butterflies*. New York: Facts on File.

Glassberg, J. 1999. *Butterflies through Binoculars: The East*. New York: Oxford University Press. 400 pp. (Covers more than 300 species in eastern North America, using color photographs in a field-guide format.)

———. 2001. *Butterflies through Binoculars: The West*. New York: Oxford University Press. 374 pp. (Includes more than 1,136 color photos, covering all the butterfly species west of central Nebraska in a field-guide format.)

———. 2017. *A Swift Guide to the Butterflies of North America*. 2nd ed. Princeton, NJ: Princeton University Press. 436 pp. (Includes all known North American species, with 3,500 color photos; probably the best-illustrated guide to the US butterflies.)

Heitzman, J. R., and J. E. Heitzman. 1996. *Butterflies and Moths of Missouri*. Jefferson City: Missouri Department of Conservation. 385 pp. (Includes photo and descriptions of about 200 species of butterflies and 150 of moths.)

Hodges, R. W. 1971. *The Moths of America North of Mexico*, Fasicle 21: Sphingoidae (Hawkmoths). London: E. W. Classey. 158 pp. (One of many technical volumes on North American moths.)

Holland, W. J. 1903. *The Moth Book*. New York: Doubleday, Doran. 479 pp. (Reprint, 1968, NY: Dover.) (The classic reference on North American moths, even if badly outdated.)

Howe, W. H. 1975. *The Butterflies of North America*. Garden City, NY: Doubleday. (Includes 2,093 paintings.)

Johnson, K. 1972. The butterflies of Nebraska. *Journal of Research on the Lepidoptera* 11: 1–64. (An annotated species list.)

Jordison, J. 1996. Jewels of the night [underwing moths]. *NEBRASKAland* 74(4): 8–19. (Includes color photographs of mounted underwing moth specimens.)

Leussler, R. A. 1972. *An Annotated List of the Butterflies of Nebraska, with the Description of a New Species*. The Mid-Continent Lepidoptera series 4(56). St. Paul, MN: John H. and Wilma L. Masters. 26 pp.

Marrone, G. M. 2002. *Field Guide to Butterflies of South Dakota*. Pierre: South Dakota Department of Game, Fish, and Parks. 175 pp. (Illustrates and describes 177 species, with range maps.)

Messenger, C. 1997. The sphinx moths of Nebraska. *Transactions of the Nebraska Academy of Sciences* 34: 89–131. (Color photographs of mounted specimens and technical descriptions of all the Nebraska species.)

Opler, P. A., A. B. Wright, and R. T. Peterson. 1994. *Peterson First Guide to Butterflies and Moths*. Boston: Houghton Mifflin. 128 pp. (Includes Peterson-style paintings and field identification information for 183 species.)

Opler, P. A., and A. M. Wright. 1999. *A Field Guide to Western Butterflies*. Boston: Houghton Mifflin. 560 pp. (Includes more than 590 species; geographic coverage extends east to central Nebraska, with Peterson-style paintings.)

Opler, P. A., and V. Maliku. 1998. *A Field Guide to Eastern Butterflies*. Boston: Houghton Mifflin. 486 pp. (Includes 524 species; geographic coverage extends west to central Nebraska, with 348 range maps and 541 Peterson-style paintings.)

Parenti, U. 1978. *The World of Butterflies and Moths: Their Life Cycle and Ecology*. New York: Putnam.

Powell, J. A., and P. A. Opler. 2009. *Moths of Western North America*. Berkeley: University of California Press. 383 pp. (About 2,500 species are described, including their anatomy, taxonomy, behavior, and life cycles, and illustrated with color photos.)

Pyle, R. M. 1981. *The National Audubon Society Field Guide to North American Butterflies*. New York: Knopf. 928 pp. (Includes more than 1,000 color photographs that illustrate all the butterflies north of Mexico.)

Ratcliffe, B. 1993. Nebraska's giant silkmoths. *NEBRASKAland* 71(3): 22–29.

Schlicht, D. W., J. C. Downey, and J. C. Nekola. 2007. *The Butterflies of Iowa*. Iowa City: University of Iowa Press. (Includes color photographs and descriptions of 118 species.)

Scott, J. A. 1992. *Butterflies of North America*. Stanford, CA: Stanford University Press. 584 pp. (Identification guide and thorough natural histories for all the North American species; too large to be an easily portable field guide but probably the best single-volume work on North American butterflies.)

Selman, C. L. 1975. *A Pictorial Key to the Hawkmoths (Lepidoptera: Sphingidae) of Eastern United States (except Florida)*. Columbus: Biological Notes No. 9, Ohio Biological Survey. 31 pp.

Spomer, S. M. 1993. Nebraska butterflies. *Museum Notes* 85. 6 pp. Lincoln: University of Nebraska State Museum. (A brief review, with photos of some of the more common species.)

Tilden, J. W., A. C. Smith, G. Christman, and B. Yung. 1986. *A Field Guide to Western Butterflies*. Roger Tory Peterson Field Guide Series. Boston: Houghton Mifflin. (Updated and replaced by Opler and Wright, 1999.)

Tuskes, P. M., M. L. Collins, and J. P. Tuttle. 1996. *The Wild Silk Moths of North America: A Natural History of the Saturniidae of the United States and Canada*. Ithaca, NY: Cornell University Press. 250 pp.

Tuttle, J. P. 2007. *The Hawk Moths of North America: A Natural History Study of the Sphingidae of the United States and Canada*. Bakersfield, CA: Wedge Entomological Research Foundation. 253 pp. (127 species)

Tveten, J., and G. Tveten. 1996. *Butterflies of Houston and Southeast Texas*. Austin: University of Texas Press. 292 pp. (Includes more than 100 species, with superb photos and descriptions of larvae and adults.)

Wagner, D. 2005. *Caterpillars of Eastern North America: A Guide to Identification and Natural History*. Princeton Field Guides. Princeton, NJ: Princeton University Press. 512 pp. (Incudes 1,200 color photos, covering nearly 700 species of butterflies and moths.)

Wright, A. B. 1993. *Peterson First Guide to Caterpillars of North America*. Boston: Houghton Mifflin: 128 pp. (Describes and illustrates 120 species of butterflies and moths, along with food plants and adult forms.)

### Dragonflies and Damselflies

Abbott, J. C. 2005. *Dragonflies and Damselflies of Texas and the South-Central United States: Texas, Louisiana, Arkansas, Oklahoma, and New Mexico*. Princeton, NJ: Princeton University

Press. (Describes 263 species, including 85 damselflies and 178 dragonflies, with keys, drawings, and range maps.)

Beckemeyer, R. J. 2002. Odonata of the Great Plains states: Patterns of distribution and diversity. *Bulletin of American Odonatology* 6(3): 49–99. https://www.odonatacentral.org/index.php/IssueAction.getFile/issue_id/128/volume_id/34/disposition/inline#page=3

Beckemeyer, R. J., and D. G. Huggins. 1997. Checklist of Kansas dragonflies. *Kansas School Naturalist* 43(2): 1–16. https://www.emporia.edu/ksn/v43n2-february1997/

Cody, J. 1996. *Wings of Paradise: The Giant Saturniid Moths*. Chapel Hill: University of North Carolina Press.

Corbett, P. S. 1963. *A Biology of Dragonflies*. Chicago: Quadrangle Books.

———. 1999. *Dragonflies: Behavior and Ecology of Odonates*. Ithaca, NY: Cornell University Press.

Dunkle, S. W. 2000. *Dragonflies through Binoculars*. New York: Oxford University Press. 266 pp. (Includes 307 species, illustrated with more than 300 color photos.)

Keech, C. F. 1934. The Odonata of Nebraska. MS thesis, Univeristy of Nebraska, Lincoln, NE. 53 pp.

Manolis, T. 2003. *Dragonflies and Damselflies of California*. Berkeley: University of California Press. 201 pp. (Illustrated with superb paintings, with 108 species shown in 40 plates.)

Molnar, D. R., and R. J. Lavigne. 1994. *The Odonata of Wyoming (Dragonflies and Damselflies)*. Laramie: University of Wyoming, Agricultural Experiment Station, Science Monograph 37R. 143 pp.

Needham, J. G., and H. Butler-Heyward. 1929. *A Handbook of the Dragonflies of North America*. Springfield, IL: Charles C. Thomas. 364 pp. (Reprint 2018 by Forgotten Books.com.)

Needham, J. G., M. J. Westfall, Jr., and M. L May. 2000. *Dragonflies of North America*. Rev. ed. Gainesville, FL: Scientific Publications. 939 pp. (Includes 350 species, with 24 colored plates and 561 illustrations.)

Nikkula, B., J. Sones, D. Stokes, and L. Stokes. 2002. *Beginner's Guide to Dragonflies*. Boston: Little, Brown. (Includes many widespread dragonflies and also some damselflies.)

Paulson, D. 2009. *Dragonflies and Damselflies of the West*. Princeton, NJ: Princeton University Press. 535 pp. (Includes 348 species, illustrated with color photos; along with Paulson [2011] probably the best available field identification book.)

———. 2011. *Dragonflies and Damselflies of the East*. Princeton, NJ: Princeton University Press. 544 pp. (Includes 336 species, illustrated with color photos; along with Paulson [2009] probably the best available field identification book.)

Sillsby, J. 2001. *Dragonflies of the World*. Washington, DC: Smithsonian Institution Press. 216 pp. (A color-illustrated summary of Odonata biology.)

Walker, E. M., and P. S. Corbet. 1953–1973. *The Odonata of Canada and Alaska*. 3 vols. Toronto, ON: University of Toronto Press. (Includes 194 species, of which 51 are Zygoptera, with 159 black-and-white plates, keys, and detailed descriptions.)

Westfall, M. J., Jr., and M. L. May. 1996. *Damselflies of North America*. Gainesville, FL: Scientific Publishers. (Includes 168 species, with identification keys; a color supplement was separately published.)

### Grasshoppers, Katydids, and Crickets

Brust, M., W. W. Hoback, and R. J. Wright. 2008. *A Synopsis of Nebraska Grasshopper Distributions*. Lincoln: University of Nebraska Extension Service. 144 pp.

Capinera, J. R., R. D. Scott, and T. J. Walker. 2004. *Field Guide to Grasshoppers, Katydids, and Crickets of the United States*. Ithaca, NY: Cornell University Press.

Capinera, J. L., and T. S. Sechrist. 1981. *Grasshoppers (Acrididae) of Colorado: Identification, Biology, and Management*. Fort Collins: Colorado State University Experiment Station Bulletin 5845.

Hagen, A. F. 1991. Distribution maps of grasshopper species in Nebraska, based on three studies. Lincoln: University of Nebraska Agricutural Experiment Station Report 16(1).

Knutson, S., G. F. Smith, and M. H. Blust. 1983. *Grasshoppers: Identifying Species of Economic Importance*. Manhattan: Kansas State University Cooperative Extension Series S-21.

McDaniel, B. 1987. *Grasshoppers of South Dakota*. Brookings: South Dakota State University Agricultural Experiment Station Bulletin TB 89.

Otte, D. 1981, 1984. *The North American Grasshoppers*. 2 vols. Cambridge, MA. Harvard University Press. Vol. 1 (Acrididae: Gomphocerinae and Acridinae), 304 pp.; vol. 2 (Oedipodinae), 378 pp.

Pfadt, R. E. 1986. *Key to the Wyoming Grasshoppers: Acrididae and Tetrigidae*. Laramie: University of Wyoming, Agricultural Experiment Station, Mimeo Circular 210. Revised by T. J. McNary (1998), modified for electronic publication by S. Schell (2004), and available at https://www.sidney. ars.usda.gov/grasshopper/extrnlpg/ ghwywest/kwgtoc.htm. (See https:// www.sidney.ars.usda.gov/grasshopper/ID_Tools/index.htm for additional grasshopper information.)

Pfadt, R. E. 2002. *Field Guide to Common Western Grasshoppers*. Laramie: Wyoming Agricultural Experiment Station Bulletin 912. 3rd ed. https:// www.sidney.ars.usda.gov/grasshopper/ID_Tools/F_Guide/index.htm

### Plant Identification and Field Guides

Barnard, I. 2014. *Field Guide to the Common Grasses of Oklahoma, Kansas, and Nebraska*. Lawrence: University Press of Kansas. 176 pp. (Includes 70 species of grasses.)

Barr, C. A., and J. Locklear. 2015. *Jewels of the Plains: Wildflowers of the Great Plains Grasslands and Hills*. Minneapolis: University of Minnesota Press. 288 pp. (Color photographs, drawings, and keys that illustrate nearly 400 species of western Canada and adjoining states, and coil bound.)

Barth, R., and N. Ratzlaff. 2004. *Field Guide to Wildflowers: Fontenelle Forest and Neale Woods Nature Centers*. Bellevue, NE: Fontenelle Nature Association. 306 pp. (Includes about 280 mostly deciduous forest trees, 25 shrubs, 10 woody vines, 32 sedges, and 2 rushes, and is coil bound.)

———. 2007. *Field Guide to Trees, Shrubs, Woody Vines, Grasses, Sedges, and Rushes: Fontenelle Forest and Neale Woods Nature Centers*. Bellevue, NE: Fontenelle Nature Association. 218 pp.

Best, K. E., J. Looman, and J. B. Campbell. 1971. *Prairie Grasses*. Pub. No. 1413. Saskatoon, SK: Canada Department of Agriculture.

Brown, L. 1979. *Grasses: An Identification Guide*. Boston: Houghton Mifflin. (Includes 135 species of the northeastern states, with 386 line drawings.)

Chadde, S. W. 2012. *Wetland Plants of the Northern Great Plains: A Complete Guide to the Wetland and Aquatic Plants of North and South Dakota, Nebraska, Eastern Montana, and Eastern Wyoming*. Charleston, SC: CreateSpace Independent Publishing Platform. 628 pp. (Includes more than 500 species.)

——. 2017. *Prairie Plants of Illinois: A Field Guide to the Prairie Grasses and Wildflowers of Illinois and the Midwest*. Charleston, SC: CreateSpace Independent Publishing Platform. 290 pp. (Includes more than 100 species, with keys and line drawings, plus county distribution maps for Illinois and the central United States.)

Christiansen, P., and M. Müller. 1999. *An Illustrated Guide to Iowa Prairie Plants*. Iowa City: University of Iowa Press. (Illustrates nearly 300 species with line drawings, including about 30 grasses.)

Denison, E. 2008. *Missouri Wildflowers*. 6th ed. Jefferson City: Missouri Department of Conservation. 280 pp. (Includes 294 species illustrated with color photographs, organized first by flower color, secondly by period of flowering, with 163 additional species described and some identification keys.)

Dunn, C. D., M. B. Stephenson, and J. Stubbendieck. 2015. *Common Grasses of Nebraska*. Lincoln: University of Nebraska Extension Service. 178 pp. (Includes maps, line drawings, and descriptions of more than 125 species of grasses and grass-like plants.)

Farrar, J. 2011. *Field Guide to Wildflowers of Nebraska and the Great Plains*. 2nd ed. Iowa City: University of Iowa Press. 384 pp. (Includes 274 species, illustrated with color photographs, organized first by flower color and then by period of flowering.)

Freeman, C. C., and E. K. Schofield. 1991. *Roadside Wildflowers of the Southern Great Plains*. Lawrence: University Press of Kansas. 280 pp. (Coverage is from southern Nebraska to the Texas Panhandle, with 253 species illustrated with color photographs.)

Great Plains Flora Association. 1977. *Atlas of the Flora of the Great Plains*. Ames: Iowa State University Press. 600 pp. (2,217 county distribution maps of all the Great Plains flora, as documented in 1986.)

——. 1986. *Flora of the Great Plains*. Lawrence: University Press of Kansas. 402 pp. (Identification keys to the 2,217 plant taxa of the Great Plains.)

Haddock, M. J. 2005. *Wildflowers and Grasses of Kansas*. Lawrence: University Press of Kansas. (Includes 323 species illustrated with color photographs, organized first by flower color and then by period of flowering.)

Hitchcock, A. S. 1935. *Manual of the Grasses of the United States*. Washington, DC: US Department of Agriculture Publication No. 200. (Reprint 1971, New York: Dover) (Includes all US species, with drawings, keys, and maps.)

Jensen, P. N. 1965. *A Key to Native and Introduced Grasses in Nebraska*. Lincoln, NE: US Department of Agriculture, Soil Conservation Service. 39 pp.

Johnson, J. R., and G. E. Larson. 1999. *Grassland Plants of South Dakota and the Northern Great Plains*. Brookings: South Dakota Agricultural Extension Station Publication B566 (rev.). 288 pp. (Includes 289 species, with color photographs; organized taxonomically by family, the coverage includes 40 grasses and 30 half-shrubs, shrubs, and trees.)

Kaul, R., B. D. Sutherland, and S. B. Rolfsmeier. 2011. *The Flora of Nebraska*. 2nd ed. Lincoln: Conservation and Survey Division, School of Natural Resources, University of Nebraska–Lincoln. (Identification keys and distribution maps to more than 2,000 species of Nebraska's vascular plants.)

Kirkpatrick, Z. 2008. *Wildflowers of the Western Plains*. Lincoln: University of Nebraska Press. 264 pp. (Includes 186 species of the shortgrass plains from Texas to Canada, including western Nebraska, with 247 color photos plus 15 line drawings and 2 monochrome photos.)

Kirt, R. R. 1995. *Prairie Plants of the Midwest: Identification and Ecology*. Champaign, IL: Stipes Publishing. 137 pp. (Includes about 100 species of grasses and forbs, illustrated with line drawings.)

Kurtz, D. 2004. *Illinois Wildflowers*. Cloudland.net Publishing. 256 pp.

Ladd, D. 1995. *Tallgrass Prairie Wildflowers*. Helena, MT: Falcon Press. (Includes 295 species, illustrated with 320 color photographs, organized first by flower color and then by period of flowering.)

Larson, G. E., and J. R. Johnson. 1999. *Plants of the Black Hills and Bear Lodge Mountains*. Brookings: South Dakota State University Agricultural Extension Station Publication B732. 608 pp. (Includes about 600 species, illustrated with color photographs and organized taxonomically by family. See also Johnson and Larson [1999]).

Müller, M. 2000. *Prairie in Your Pocket*. Iowa City: University of Iowa Press.

(Fold-out poster of 114 prairie plants illustrated with paintings.)

*Pasture and Range Plants*. 1963. Bartlesville, OK: Phillips Petroleum Company. 176 pp. (Includes paintings of 70 species of grasses and more than 80 forbs).

Petrides, G. A. 1958. *A Field Guide to Trees and Shrubs*. Boston: Houghton Mifflin.

Richett, H. W. 1973. *Wildflowers of the United States: The Central Mountains and Plains*. Vol. 6, parts 1–3. New York: New York Botanical Gardens and McGraw-Hill.

Runkel, S. T., and D. M. Roosa. 2010. *Wildflowers of the Tallgrass Prairie: The Upper Midwest*. 2nd ed. Iowa City: University of Iowa Press. 279 pp. (Includes 273 species, organized alphabetically and illustrated with color photographs.)

Stephens, H. A. 1969. *Trees, Shrubs, and Woody Vines of Kansas*. Lawrence: University Press of Kansas.

Strickler, D. 1986. *Prairie Wildflowers*. Columbia Falls, MT: Flower Press. 80 pp.

Stubbendieck, J., and K. L. Kottas. 2005. *Common Grasses of Nebraska*. Lincoln: University of Nebraska–Lincoln, Institute of Agriculture and Natural Resources, University of Nebraska Extension Bulletin EC05 170. 121 pp. http://digitalcommons.unl.edu/extensionhist/4786/ (Covers 180 species of grasses and 7 sedges.)

———. 2007. *Common Forbs and Shrubs of Nebraska*. Lincoln: University of Nebraska–Lincoln, Institue of Agriculture and Natural Resources, University of Nebraska Extension Bulletin EC-118. 178 pp. http://digitalcommons.unl.edu/extensionhist/4799/ (Covers 117 species of forbs, 17 shrubs, and 4 cacti.)

Stubbendieck, J., J. T. Nichols, and K. K. Roberts. 1985. *Nebraska Range and Pasture Grasses.* University of Nebraska–Lincoln, Nebraska Cooperative Extension Service Circular 85-170. 75 pp. http://digitalcommons.unl.edu/extensionhist/4904/

Stubbendieck, J. L., G. Y. Frissoe, and M. R. Bolick. 1995. *Weeds of Nebraska and the Great Plains.* 2nd ed. Lincoln: Nebraska Department of Agriculture. (Includes 250 species, illustrated photographically and organized taxonomically by family.)

Stubbendieck, J. L., S. L. Hatch, and L. M. Landholt. 2003. *North American Wildland Plants.* 6th ed. Lincoln: University of Nebraska Press. (Illustrated with line drawings plus range maps; about 200 species of grassland plants are described, including grasses, forbs, shrubs, and small trees. Prior editions of this book were titled *North American Range Plants.*)

Sutherland, D. M. 1975. A vegetative key to Nebraska grasses. Pp. 283–316, in *Prairie, a Multiple View* (M. K. Wali, ed.). Grand Forks: University of North Dakota Press.

US Department of Agriculture. 1970. *Selected Weeds of the United States.* Washington, DC: Government Printing Office. (Reprint 1971 as *Common Weeds of the United States*, New York: Dover.)

Van Bruggen, T. 1983. *Wildflowers, Grasses, and Other Plants of the Northern Plains and Black Hills.* 3rd ed. Interior, SD: Badlands Natural History Association. 96 pp. (Includes color photos and short descriptions of 118 forbs and shrubs.)

Vance, F. R., J. R. Jowsey, J. S. McLean, and F. A. Switzer. 1999. *Wildflowers of the Northern Great Plains.* 3rd ed. Minneapolis: University of Minnesota Press. 384 pp. (Illustrates nearly 400 species of the Canadian and northern US plains, with color photos and drawings.)

*Weeds of the North Central States.* 1981. Illinois Agricultural Extension Bulletin 772, University of Illinois at Urbana-Champaign. 303 pp. (Includes more than 230 species.)

Wilcox, E. M., G. K. K. Link, and V. W. Pool. 1915. *A Handbook of Nebraska Grasses, with Illustrated Keys for Their Identification, Together with a General Account of Their Structure and Economic Importance.* Lincoln: University of Nebraska, Agricultural Experiment Station of Nebraska. 120 pp. http://digitalcommons.unl.edu/extensionhist/424/ (Keys and line drawings of most Nebraska grasses.)

9 781609 621315